2

BISS MONOGRAPHS
MONOGRAPHS OF THE
BREMEN INSTITUTE
OF SAFE SYSTEMS

M. Gogolla, H.-J. Kreowski, B. Krieg-Brückner,
J. Peleska, B.-H. Schlingloff, H. Szczerbicka (Series Editors)

REPRESENTATIONS, HIERARCHIES AND GRAPHS OF INSTITUTIONS

Till Mossakowski

Revised Version, December 2001

Dissertation
zur Erlangung des Grades eines Doktors der
Ingenieurswissenschaften – Dr.-Ing. –

Vorgelegt im Fachbereich 3 (Mathematik & Informatik)
der Universität Bremen
im Juli 1996

Die Deutsche Bibliothek – CIP-Einheitsaufnahme

Mossakowski, Till:
Representations, Hierarchies and Graphs of Institutions /
Till Mossakowski. - Berlin : Logos-Verl., 2002

   (BISS monographs ; Bd. 2)
   Zugl.: Bremen, Univ., Diss., 2001
   ISBN 3-89722-831-9

©Copyright Logos Verlag Berlin 2002
Alle Rechte vorbehalten.

ISBN 3-89722-831-9
ISSN 1435-8611

Logos Verlag Berlin
Comeniushof, Gubener Str. 47,
10243 Berlin
Tel.: +49 030 42 85 10 90
Fax: +49 030 42 85 10 92
INTERNET: http://www.logos-verlag.de

Gutachter:
Prof. Dr. Hans-Jörg Kreowski, Bremen
Prof. Dr. Andrzej Tarlecki, Warschau

Datum des Promotionskolloquiums:
26.08.1996

Editorial

When the ADJ-group with Goguen, Thatcher, Wagner, and Wright introduced the concept of algebraic specification about 30 years ago, they initiated a very successful and still flourishing new area. At the beginning, there was the simple and brilliant idea to model abstract data types by equations over many-sorted signatures as syntactic descriptions with initial algebras as semantics. But soon after this "big bang" the universe of algebraic specification has expanded amazingly in size and variety. On the syntactic level, one encounters hundreds of concepts in dozens of variants including conditional equations, universal Horn-clauses, first-order formulas, subsorts, hidden sorts and operations, generators, derivors, enrichments, extensions, parameterizations, parameter passing, modules, etc. And on the semantic level, initial algebras have got many companions like arbitrary models, term-generated models, terminal algebras, based algebras, partial algebras, reducts, functors, free functors, etc. Not to forget, accompanying methods have been developed like term rewriting, theorem proving, stepwise refinement, implementation, behavioural specification, etc.

Nearly every researcher in the field proposed new concepts, but ignored usually the relationship to other existing ones. As a result, the area of algebraic specification got somewhat messy so that the call for order has come up louder and louder. For a start, Burstall and Goguen suggested the notion of institutions to show that most of the concepts around form thousands of institutions and that some of the typical considerations are independent of the underlying institution. But how are all those institutions which arise in the framework of algebraic specification related to each other?

Most of Till Mossakowski's research work has been devoted to this complex topic. His thesis reports on these efforts presenting a collection of enlightening results. He has detected some significant order in the chaos and grouped a great number of institutions into galaxies of equivalent expressiveness. The thesis provides a very careful study and classification of many institutions that are related to partiality in one way or the other. Moreover, mappings between these institutions are constructed that allow to compare their expressive powers. As a result, five different levels are identified with growing expressiveness, and the institutions on the same level are shown to be equivalent. The comparison of institutions does not only concern the class of models and in this way the loose semantics, but is also exended to free functors and parameterization as well as to proof-theoretical aspects. The careful and systematic study of a considerable number of institutions and their relationship to each other sheds much light on the various ways of specification and yields a deep insight into the expressive power of logical systems.

The reader may enjoy an excellent text on a key problem of formal system specification that can be the basis for further breakthroughs in the future research on algebraic specification.

Bremen, 17 December 2001　　　　　　　　　　　　　　　Hans-Jörg Kreowski

Summary

For the specification of abstract data types, quite a number of logical systems have been developed. In this work, we will try to give an overview over this variety. As a prerequisite, we first study notions of *representation* and embedding between logical systems, which are formalized as *institutions* here. Different kinds of representations will lead to a looser or tighter connection of the institutions, with more or less good possibilities of faithfully embedding the semantics and of re-using proof support.

In the second part, we then perform a detailed "empirical" study of the relations among various well-known institutions of total, order-sorted and partial algebras and first-order structures (all with Horn style, i.e. universally quantified conditional, axioms). We thus obtain a *graph* of institutions, with different kinds of edges according to the different kinds of representations between institutions studied in the first part.

We also prove some separation results, leading to a *hierarchy* of institutions, which in turn naturally leads to five subgraphs of the above graph of institutions. They correspond to five different levels of expressiveness in the hierarchy, which can be characterized by different kinds of conditional generation principles.

We introduce a systematic notation for institutions of total, order-sorted and partial algebras and first-order structures. The notation closely follows the combination of features that are present in the respective institution. This raises the question whether these combinations of features can be made mathematically precise in some way. In the third part, we therefore study the combination of institutions with the help of so-called parchments (which are certain algebraic presentations of institutions) and parchment morphisms.

The present book is a revised version of the author's thesis, where a number of mathematical problems (pointed out by Andrzej Tarlecki) and a number of misuses of the English language (pointed out by Bernd Krieg-Brückner) have been corrected. Also, the syntax of specifications has been adopted to that of the recently developed Common Algebraic Specification Language CASL [CASL, Mos97].

Acknowledgments

I want to thank Hans-Jörg Kreowski for teaching me algebraic specifications, for giving me the freedom to find my own line of research interest and for always being open for questions and discussions. Moreover, I thank all the members of the research group "Theoretical computer science" in Bremen for the really good working atmosphere.

Thanks also to Andrzej Tarlecki. I learned much from him in numerous discussions on the subject. Though some of the approaches presented in this thesis might not follow his view, on the main topics, and on my personal ideas and views of the subject, he had a strong influence. In particular, he had the idea to use monads for describing various kinds of institution representations (see chapter 4).

Thanks also to (in reverse alphabetical order) Uwe Wolter, Burkhard Wolff, Wiesiek Pawłowski, Amilcar Sernadas, Pippo Scollo, Horst Reichel, Pepe Meseguer, Bernd Krieg-Brückner, Jo Goguen, Maura Cerioli, Valentin Antimirov and the FLIRTS (Formalism-Logic-Institution Relating, Translating and Structuring) community for various discussions.

This work has been supported by a PhD scholarship of the Studienstiftung des deutschen Volkes and partially by the ESPRIT Basic Research Working Group COMPASS II.

Contents

1 **Introduction**   1
- 1.1 Data types over an arbitrary but fixed signature 2
- 1.2 Modular data types: moving between signatures 3
- 1.3 Specification of abstract data types in an arbitrary but fixed logical system 5
- 1.4 Moving between logical systems 6
- 1.5 The problem to choose a not arbitrary, but fixed category of logical systems 7
- 1.6 Moving between categories of logical systems 9
- 1.7 The role of model morphisms, initial and free models 10
- 1.8 Structure of the thesis 11

2 **Some categories of logical systems**   13
- 2.1 Specification frames 13
- 2.2 Institutions 14
- 2.3 Composable signatures and amalgamation 17
- 2.4 Entailment systems 23
- 2.5 Logics 24
- 2.6 Transporting logical structure along maps 24
- 2.7 Liberality 26

3 A variety of institutions of total, partial and order-sorted algebras   27

   3.1   Relational Partial Conditional Existence-Equational Logic   .   27

   3.2   Restrictions of $RP(R \stackrel{e}{=} \Rightarrow R \stackrel{e}{=})$ 32

   3.3   Further restrictions of $RP(R \stackrel{e}{=} \Rightarrow R \stackrel{e}{=})$ 33

   3.4   Some restrictions with special interpretation 35

   3.5   Limit theories 36

   3.6   Some algebraic notions and propositions 36

   3.7   Left exact sketches 39

   3.8   Order-sorted algebra with sort constraints 40

   3.9   Composable signatures and amalgamation 43

   3.10 Liberality of some institutions 46

4 Different types of arrow between logical systems   49

   4.1   The theory monad and simple representations of institutions   50

   4.2   The conjunctive monad and conjunctive representations of institutions 52

   4.3   The presentation extension monad and weak representations of institutions 53

   4.4   Semantical consequence in **derive!**$(\mathcal{F})$ and **derive!**$(\mathbf{Pres}(\mathcal{I}))$ . . 60

   4.5   The model class monad and semi-representations of institutions . 62

   4.6   Summary 64

   4.7   Subinstitutions and equivalent expressiveness of institutions   . 66

   4.8   Is our notion of equivalence strong enough? An example 69

5 Five graphs of institutions   73

   5.1   Equivalences among various institutions at level 5 74

   5.2   Categorical intuitionistic type theory 95

   5.3   Equivalences at level 4, and embeddings to level 5 96

   5.4   Equivalences at level 3, and embeddings to level 4 96

   5.5   Equivalences at level 2, and embeddings to level 3 102

5.6		Level 1, and embeddings to level 2	105
5.7		Rewriting logic	105

6 Hierarchy theorems 107

6.1		A model theoretic hierarchy theorem	109
	6.1.1	Level 5 versus level 4: Partial Conditional Logic and Horn Clause Logic	109
	6.1.2	Level 4 versus level 3: Horn Clause Logic and Conditional Equational Logic	111
	6.1.3	Level 3 versus level 2: Conditional Equational Logic and Partial Equational Logic	112
	6.1.4	Is there a subhierarchy within level 2?	114
	6.1.5	Level 2 versus level 1: Partial Equational Logic and Total Equational Logic	115
6.2		Initial semantics	117
6.3		Properties of parameterized ADTs	118
6.4		Example PADTs and a hierarchy theorem	124
6.5		Locating bounded stacks in the hierarchy	129
6.6		Summary	130

7 Equivalence and Difference of Institutions 133

7.1		Conditional Equational Logic and Based Algebras	134
7.2		The *PART* construction	136
7.3		Measuring the difference of institutions: categorical retractive representations	138
7.4		*PART* is categorical retractive	141
7.5		Other categorical retractive representations	143
	7.5.1	Level 5 versus level 4	144
	7.5.2	Level 4 versus level 3	145
	7.5.3	Level 3 versus level 2	145
	7.5.4	Level 2 versus level 1	146
7.6		Specifiability of representations	147

7.7	Specifiability of *PART*	150
7.8	Summary	156

8 Parchments — a device for combining logics   159

8.1	Why institution morphisms do not suffice	160
8.2	Parchments	162
8.3	Parchment morphisms	165
8.4	Putting parchments together using limits	167
8.5	Comparison with related work	174

9 Conclusion   177

A Some category-theoretic preliminaries   181

A.1	Monads and Kleisli categories	181
A.2	Multiple pushouts, multiple pullbacks and amalgamation	182
A.3	Locally finitely presentable categories	183
A.4	Effective equivalence relations	184
A.5	Foundational issues	185

Index   197

Chapter 1

Introduction

> "Recent work in theoretical computer science uses many different logical systems. Perhaps most popular are the many variants of first and higher order logic found in current generation theorem provers. But also popular are equational logic, as used to study abstract data types, and Horn clause logic, as used in 'logic programming', e. g., Prolog. More exotic logical systems such as temporal logic, second order polymorphic lambda calculus, dynamic logic, order-sorted logic, modal logic, continuous algebra, infinitary logic, intuitionistic higher order type theory, and intensional logic have been proposed to handle problems such as concurrency, overloading, exceptions, nontermination, program construction and natural language. However, it seems apparent that many general results used in the applications are actually *completely independent* of what underlying logic is chosen." *J. A. Goguen, R. Burstall* [GB92]

The complexity of software systems has grown in a way that has made methodologies and tools that guide and support the development process become more and more important. A crucial problem is to translate a description or solution of a problem from some informal to a formal language. There is a large gap between an informal requirement and a program (which is necessarily written in a formal language). Now the methodology of algebraic specification of abstract data types provides a formal language with mathematical semantics already at the level of requirement specification, so the gap between the informal requirement and the first formal document becomes smaller. Moreover, once a specification has been formalized, the development process can be supported by semi-automatic tools, since the formal semantics allows to formally define the notion of correctness of implementations, to derive consequences which are implied by a specification, and to stepwise transform requirement specifications into design specifications and implementations.

Further, the method of algebraic specification follows the principle not to construct abstract data types explicitly, but rather to describe their abstract properties. This allows to concentrate on the essential requirements, which potentially may be refined further, and delay the complete design and the implementation until later.

There is a variety of specification languages for the formal specification and development of correct software systems. Some examples are CLEAR [BG77], ASL [Wir86], Larch [GH86], ACT ONE [EM85], Extended ML [ST90, ST86] and OBJ3 [GW88]. These specification languages differ in purpose, expressiveness, level of abstraction (requirement, design, implementation), notation, available tools etc. Some ingredients of these languages vary, some are the same. Moreover, all of them are based on some kind of logical system.

I now want to introduce various levels at which logical systems can be considered, and at each level indicate how this thesis contributes to the theory of logical systems.

1.1 Data types over an arbitrary but fixed signature

A signature provides names for individual program components (for example, types, also called sorts, and functions, or operations). It describes which ingredients a data type should have. Consider the problem of sorting a list of natural numbers:

>   **spec Sorting** =
>   **sorts**   nat, list
>   **ops**   zero : nat;
>   succ : nat $\to$ nat;
>   empty : $\to$ list;
>   add : nat $\times$ list $\to$ list;
>   sort : list $\to$ list
>   **preds**   _ $\leq$ _ : nat $\times$ nat

A model (or data type) over this signature can be described within the usual (semi-formal) language of mathematics:

>   **algebra** A =
>   **Carriers**
> $|A|_{\mathtt{nat}} = \mathbf{N}$
> $|A|_{\mathtt{list}} = \mathbf{N}^*$
>   **Functions**

$\text{zero}_A = 0$
$\text{succ}_A(n) = n + 1$
$\text{empty}_A = \lambda$
$\text{add}_A(x, l) = x \cdot l$
$\text{sort}_A(l) =$ the permutation of l which is in ascending order

Predicates

$\_ \leq \_{}_A = \{\, (m, n) \mid m \leq n \,\}$

Any implementation of sorting of lists of natural numbers yields such a model. At the stage of requirement specification, we do not want to implement a data type yet, but rather describe it abstractly by its properties. This leads to an abstract data type, since for the properties to hold or not to hold, the concrete representation of the data is inessential. For example, for the ordering on the natural numbers, the following axioms hold:

$$\forall x : \text{nat} .\ \text{zero} \leq \text{succ}(x)$$

$$\forall x, y : \text{nat} .\ x \leq y \Rightarrow \text{succ}(x) \leq \text{succ}(y)$$

Usually specifications are considered to be *loose*, that is, we talk about the realm of all models of a given specification (which may be narrowed be refining the specification). But there are situations where it is also useful to have a canonical abstract data type as semantics, given by the *initial* models of a specification.

From a given set of axioms, we can derive its *consequences* as theorems, using some proof calculus or theorem prover. Thus we may check if the logical consequences of a given specification are consistent with our informal specification. Of course, the proof calculus depends on the semantics. For example, loose semantics has a calculus different from initial semantics.

1.2 Modular data types: moving between signatures

The mathematical theory of abstract data types of an arbitrary but fixed signature is developed mainly in two branches of mathematics: universal algebra and mathematical logic. But for computer science, this view does not suffice. To be able to specify data types in a modular way, we need a notion of morphism between signatures, which allows to change signatures smoothly.

To be able to specify the list sorting example correctly, we need to enrich it by some auxiliary operations, which may be hidden later. Moreover, we do not want to consider just lists over natural numbers, but over an arbitrary element type, which leads to a parameterized specification:

```
spec Nat =
    sorts   nat
    ops     zero : nat;
            succ : nat → nat
    preds   _ ≤ _ : nat × nat
    ∀x : nat . zero ≤ succ(x)
    ∀x, y : nat . x ≤ y ⇒ succ(x) ≤ succ(y)

spec Elem =
    sorts   elem
    preds   _ ≤ _, _noteq_ : elem × elem

spec List[Elem] given Nat =
    sorts   list
    ops     empty : list;
            add : elem × list → list;
            _ + + _ : list × list → list;
            count : elem × list → nat
    ∀l : list . empty + + l = l
    ∀e : elem, l, l' : list . add(e, l) + + l' = add(e, l + + l')
    ∀e : elem . count(e, empty) = zero
    ∀e, e' : elem, l : list . e = e' ⇒ count(e, add(e', l)) = succ(count(e, l))
    ∀e, e' : elem, l : list . e noteq e' ⇒ count(e, add(e', l)) = count(e, l)

spec SortList[Elem] given Nat =   List[Elem] then
    ops     sort : list → list
            ∀e, e' : elem, l, l', l'' : list .
                sort(l) = l' + + add(e, add(e', empty)) + + l'' ⇒ e ≤ e'
    ∀l : list, e : elem . count(e, l) = count(e, sort(l))
```

Usually one would specify the predicate _noteq_ to be the complement of equality, but in order to get positive conditional specifications, which are studied in this work, we leave this out here.

In the further development, we may want to combine this specification with others while sharing the specification of natural numbers, or way may want to rename something. All this can be done using signature morphisms, which typically induce translations of sentences and reductions of models (where model reduction is done in the direction *opposite* to signature translation). Sentence and model translation are linked by the important *satisfaction condition* [GB92] stating that *satisfaction is invariant under moving between signatures along signature morphisms*.

To make the above example work properly, we have to add an *initiality constraint* which ensures that Nat is interpreted by the natural numbers, and a *free generating constraint* which ensures that any model of List is generated freely over its Elem-reduct.

1.3 Specification of abstract data types in an arbitrary but fixed logical system

From the previous sections it becomes clear that a specification language both contains

1. constructs for specification-in-the-small, that is, specifying properties of individual program components, and

2. specification-building operators (like extension, union, renaming, hiding) for constructing more complex specifications from simpler ones.

With Goguen's and Burstall's institutions [GB92] as a formalization of the notion of logical system, Sannella and Tarlecki [ST88a] show that it becomes possible to separate a specification language into the following to levels:

1. choose a particular institution, providing notions of signature, signature morphism, model, and satisfaction,

2. choose a set of specification-building operators, which can be defined in an institution independent way.

The second point is an important issue on its own, but not studied in this thesis. Thus, if we assume a canonical choice of specification-building operators to be given (see, for example, [ST88a]), then we only need to supply an institution to get a specification language.

Among others, the following institutions are examined in this work:

Relational Partial Conditional Existence-Equational Logic (abbreviated $RP(R \stackrel{e}{=\!\!\Rightarrow} R \stackrel{e}{=})$) has as signatures many-sorted signatures including symbols for partial operations (indicated by the P) and relations (or predicates, indicated by the R). Models are many-sorted algebras (or, more precisely, partial algebraic systems, since there are relations). Signature morphisms have to map sorts to sorts, in a way that this is compatible with the mapping of the other components. Reducts are defined by renaming the components of a model along the signature morphism. Sentences are universally quantified conditional (indicated by $\Rightarrow$) existence-equations (indicated by $\stackrel{e}{=}$) mixed with relation applications (indicated by R); see the books of Burmeister [Bur86] and Reichel [Rei87].

$R(R =\!\!\Rightarrow \exists!R =)$ is a fragment of many-sorted first-order logic described by Coste [Cos79] which allows a restricted use of unique-existential quantification. Sentences are universally quantified conditionals, where, as indicated by the expression in brackets, the premise consists of a conjunction of equations and/or relations applications, and the conclusion is a unique-existentially quantified such conjunction.

$COS(=:\Rightarrow=:)$ is the institution of coherent order sorted signatures, algebras and theories introduced by Goguen and Meseguer [GM92] with conditional sort constraints [GJM85, Yan93].

$LESKETCH$, the institution of left exact sketches [Gra87, BW85] has graphs as signatures, which are interpreted in the category of sets. Sentences allow to state commutativity of diagrams and properties of products (and other limits in the sense of category theory).

In general, the choice of the institution determines whether we just have total operations available, or also predicates (as in the above sorting example) or partial operations, which notions of built-in equality are available, which kinds of logical connectives and quantifiers there are, if the logic is first-order or higher-order, and so on. But also modal, temporal, object behaviour, process and relevance logics and even programming languages can be considered to be institutions.

In this thesis, I can cover only a small part of this variety of institutions. I concentrate on institutions of total, order-sorted and partial algebras; they are introduced in chapter 3. All these institutions are liberal, which means that initial models (and free constructions) exist and can serve as canonical semantics for (parameterized) specifications. Perhaps even more important is that these institutions allow us to use theorem proving tools like conditional term rewriting or paramodulation, which are not as complex as tools for other institutions. Even within this restricted realm of institutions, a systematic study of them requires some effort. Of course, the next step could be to extend the systematic study to other institutions.[1] But before doing this, I find it more useful to develop meta-theoretical means to systematically combine concepts from different institutions, see chapter 8. Otherwise, one can easily get stuck in the overwhelming manifold of existing formalisms.

1.4 Moving between logical systems

There is a large number of institutions (some of them sketched in the previous section), which differ in purpose, expressiveness, personal taste of its user, and so on. It seems useless to try to construct one universal institution serving all purposes, because such a thing would be very clumsy, and sooner or later, due to the combination of too many concepts, it would be difficult or even impossible to develop a clear intuitive understanding, a mathematical semantics, tool support etc. On the other hand, it is unsatisfactory that specifications written

---

[1] Actually, the results of this next step can be found in [Mosb].

in different institutions are totally unrelated. Typically, each team uses its own institution... To be able to relate them, we need a notion of representation (or encoding) of one institution into another. Technically, this leads to a *category of institutions and representations*.

With this, one can compare the *expressiveness* of institutions. In chapter 5, a certain number of institutions are embedded into each other, while chapter 6 proves some separation results. The outcome is a *graph of institutions and representations*, which represents some objects and morphisms in the category of institutions and representations. It can be used, for a given specification problem, to find an institution that is strong enough to express the problem but weak enough to have good support available (theorem provers, specification language frameworks).

Another topic are multi-paradigm specification languages, advocated by Astesiano and Cerioli [AC94], Sannella and Tarlecki [ST88b, Tar96a, Tar99], Diaconescu [Dia01] and others, which allow to write heterogeneous specifications with components written in different institutions. One can select a representation from the institution graph and use it to get a specification-building operator that translates a specification from one institution to the other [Tar96a]. An important question is whether these new specification-building operators commute with the old ones, that is, whether institution representations preserve specification-building operators [Cer93].

1.5 The problem to choose a not arbitrary, but fixed category of logical systems

Many different categories of logical systems are described in the literature: institutions [GB92] with maps of institutions [Mes89], specification frames [EPO89], institutions with simulations [AC92], entailment systems with maps [Mes89], proof calculi with maps [Mes89], logics with maps [Mes89], pre-institutions with transformations [SS92], institutional frames [Wol95], π- and τ-institutions [FS88, SS95], foundations [Poi89], galleries [May85], context institutions [Paw96], parchments [GB85] with various notions of morphisms [ST, Mos96c, MTP98] etc. Some of these were related by Maura Cerioli in her thesis [Cer93]. Thus, at the meta-level, we have to choose among different types of logical system (resp. categories of logical systems). Even if the objects of such a category remain fixed, often there are several reasonable choices what the morphisms should be.

But if we do not follow some neat structuring principle in the meta theory (i. e. types of logical system), there is the danger to end up in the same Babylonian realm of different languages as in the object theories (i. e. different logical

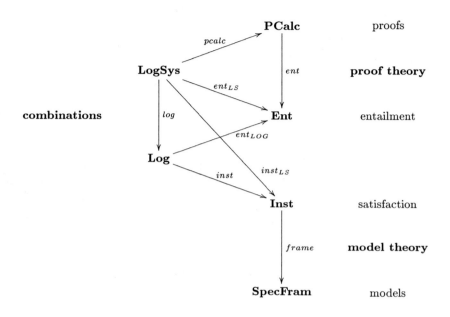

Figure 1.1: Different categories of logical systems and representations, related by some forgetful functors

systems). After all, the meta theory should help us to structure and compare the variety of logical systems, and not to produce the same diversity of notions! Probably, there are much fewer arguments for introducing a new type of logical system than for introducing a new logical system; even if there are some important arguments that show that there cannot be "the" category of logical systems serving all purposes equally well.

The main argument for having several types of logical system is the distinction between proof theory and model theory, and the question which proof-theoretical and/or model-theoretical logical components are given. *Specification frames* just consider specifications and models. *Institutions* split specifications into signatures and sentences, and a satisfaction relation (between models and sentences) is provided. *Parchments* further specify an abstract syntax of sentences, along which satisfaction can be defined inductively using an algebra of term evaluators. *Context institutions* allow to deal with variable contexts and substitutions.

On the other hand, *entailment systems* provide a ("syntactical") entailment relation between sets of sentences and sentences, while *proof calculi* also provide proofs for such entailment. *Logics* and *logical systems* combine both model theoretic and proof theoretic components. In this thesis, I will concentrate on model theory, but also consider how proof theory is affected by representations.

A second argument for having different *categories* of logical systems is the purpose of the morphisms. Tarlecki [Tar96a] distinguishes the following: One purpose is, as said above, the representation of one logical system in another. But there is also the orthogonal purpose of building one logical system above another, which is captured by the notion of *institution morphism* [GB92]. This is the basis for the combination of logical systems and study of interaction of concepts in a systematic way. Some work in this direction is done in chapter 8 and applied to some of the institutions of the institution graph. Moreover, implementations typically go along *institution semi-morphisms* [ST88b, Tar96a], which relate, for example, a high-level specification language and a programming language.

A third argument is the observation that different degrees of more or less good representation of one logical system in another are possible. In chapter 4, it is studied, which subclass of representations should be called embeddings, or come at least close to embeddings. In the same chapter I argue that many examples require some intermediate notion between institution representation and specification frame representation. This has to do with the distinction of signatures and sentences in an institution. This distinction is necessary (using just specification frames abandons the notion of sentence and of theorem proving), but there are many choices for what should be included into the signature part, and what into the sentences. Therefore there is a need for types of arrow that may "mix up" this distinction. In chapter 4, I show that these different types of arrow can be generated by one basic type of arrow and monadic constructions on categories of logical systems, with the effect of automatically having functors relating the new categories of logical systems with the old ones.

1.6 Moving between categories of logical systems

The alert reader may have expected how the story now continues: after considering an arbitrary, but fixed thing, we next consider moving between such things via morphisms. After all, this is the lesson of category theory: do not consider just the objects themselves, but also the relations between them. This idea is now has been iterated over various levels in the previous sections. But don't bother, we will reach the end of the story soon!

Different categories of logical systems can be related via functors. Cerioli and

Meseguer [CM93] describe two kinds of such functors:

1. Forgetful functors, which forget some part of the logical structure of a logical system. For example, an institution yields a (generally very poor) specification frame by just forgetting that there are sentences and satisfaction.

2. Functors which add some part of the logical structure in some canonical way. For example, I show in chapter 4 how to add sentences and satisfaction to a specification frame. Typically, these functors are *right* adjoints to the forgetful functors.

Given such an adjoint pair of functors relating different categories of logical systems, Cerioli and Meseguer show how missing logical components, such as entailment system, proof theory or model theory can be borrowed from another logical system (which is of a richer type) by transporting these logical components along a representation map. In chapter 4, this borrowing idea is applied to sentences as well.

Apart from these, there are other functors between logical systems that are close to forgetful functors: they not only forget things, but also construct things out of other things, for example, the construction of an institution out of a parchment. It is important to relate each new category of logical systems to existing categories of logical systems in order to have a chance to relate and unify the theories that are built around those categories.

All six levels of abstraction introduced in the preceeding six sections have their own right and are studied in the literature. But they should be kept together in some way. Studying the general theory without concrete examples may lead to theoretical notions without any content, and sticking to concrete examples may result in either a narrow horizon or a loss of overview.

1.7 The role of model morphisms, initial and free models

Model morphisms play an important role in (universal) algebra. In general, they are defined in such a way that they preserve the structure of models. One can revert this and say that model categories (which are essentially determined by the morphisms) implicitly define the structure that is around in the models. Hilberdink [Hil95] extracts the syntax of terms from model categories.

On the other hand, model morphisms are less important in mathematical logic and the theory of logical systems, because they do not preserve satisfaction, which makes them somewhat unrelated to the logical structure. Thus, the subject of chapter 8, combination of institutions, is entirely independent of model morphisms. But this is not true for other chapters. For the definition of embedding representations in chapter 4, it is essential to have a notion of model isomorphism, since otherwise (counting only identities as morphisms, which is always possible) many useful embeddings would be excluded.

Initial models provide canonical models for specifications. In many institutions, these are characterized as being term-generated and not confusing terms unless this is implied by the specification. In a similar manner, free functor semantics yields canonical parameterized abstract data types. But also within loose semantics, these concepts can be used via initiality and free generating constraints [EM90], which state that some part of the model shall be initial or the result of a free functor. The definition of initiality and freeness is heavily based on the whole class of model morphisms. Since all the institutions examined in this work are liberal (which means the initial and free models exist), it is also useful to consider how initiality and freeness is preserved by representations (see chapter 7 on categorical retractive simulations).

Another important use of model morphisms concerns negative results about the existence of embeddings. For these results, the category-theoretic classification theorems of Adámek and Rosický [AR94] can be used. These category-theoretic results are reflected by similar separation results which show how the different levels of expressiveness behave with respect to free constructions, see chapter 6.

If one is not interested in all model morphisms, but just in the isomorphisms (perhaps with a requirement that these should preserve satisfaction), then the category-theoretic classification results of [AR94] do not longer apply (though I conjecture that many of the general classification results still remain true). One might then use the results of classification and stability theory [Bal88, She78], which are branches of model theory. But this is beyond the scope of this thesis.

1.8 Structure of the thesis

Chapter 2 recalls the categories of logical systems that are needed for the further study. The emphasis is laid on the model-theoretic types of logical system: institutions and specification frames. Moreover, the question of combination of theories is treated both on the syntactical (composable signatures and theories) and on the semantical level (amalgamation). Concerning the preservation of composable theories and amalgamation, some new definitions and theorems arise.

In chapter 3, a variety of institutions of total, partial and order-sorted algebras, and some algebraic theorems are recalled from the literature.

Chapter 4 introduces four different types of representation between institutions, three of which are new. These types of arrows can be generated by a general categorical construction, which, as a by-product, leads to the relation of the different resulting categories through (adjoint) functors. Finally, a subclass of representations (of each type) is singled out to serve as *embeddings*. This is the central notion for the comparison of expressive power.

Chapter 5 uses these types of representation to systematically embed the institutions from chapter 3 into each other. Five levels of expressiveness emerge.

In Chapter 6, these five levels are separated from each other, first by model-theoretic properties, and then by properties of parameterized abstract data types.

Chapter 7 studies constructions that interpret theories and models in an institution to denote more complex objects. For example, the *PART*-construction [Kre87] interprets total based algebras as partial algebras, or ET logic [MSS90] interprets a predicate symbol as a typing relation. These interpretations are captured by the notion of categorical retractive representation, which is a weakening of the notion of embedding. With this, two levels of proof-theoretic complexity emerge.

Finally, Chapter 8 leaves the path I have been following so far and switches over to a new perspective: the combination of logical systems. While this is formally almost entirely orthogonal to the other chapters, it is motivated conceptually by the overwhelming possibilities to combine logical systems. The systematic notation for institutions introduced in chapter 3 allows to structure the realm of institutions a bit. Chapter 8 tries to give a semantic foundation to this notation.

Chapter 9 concludes the thesis, and appendix A contains some category theoretical prerequisites.

Chapter 2

Some categories of logical systems

> "What unique contributions to the broader area of software foundations and formal methods do these conceptual developments place us in a position to make? Probably many, but we believe that a particularly pressing problem that they could solve is the serious need for *formal interoperability* among different formal methods and specification formalisms, that is, the capacity to move in a mathematical rigorous way across the different formalizations of a system, and to use in a rigorously integrated way the different tools supporting such formalizations." *N. Martí-Oliet* and *J. Meseguer* [MOM95]

In this chapter, some categories of logical systems, namely specifications frames, institutions, entailment systems, and logics (each with representations as morphisms) are recalled from the literature and related by some functors.

2.1 Specification frames

Specification frames by Ehrig, Pepper and Orejas [EPO89] formalize abstract specifications and models of specifications, while there are no notions of sentence and satisfaction. We here use "theory" instead of "specification" because the number of axioms may be infinite, while a specification is generally to be considered finite. (On the other hand, this may confuse those who think of theories as being closed under semantical consequence – this is *not* required here.)

Definition 2.1.1 A *specification frame* $\mathcal{F} = (\mathbf{Th}, \mathbf{Mod})$ consists of

1. a category **Th** of theories and
2. a functor **Mod**: $(\mathbf{Th})^{op} \longrightarrow \mathbf{CAT}$ [1] giving the category of *models* of a theory. □

We write $M'|_\sigma$ (the σ-*reduct* of M' under σ) for $\mathbf{Mod}(\sigma)(M')$. If $M = M'|_\sigma$, M' is called a σ-*expansion* of M.

Note that models are translated in the direction *opposite* to the direction of theory morphisms. If one tries to expand models *along* with theory morphisms, there is the difficulty that an extension of a theory causes the need to supply some additional components in the models. And it is not clear what these additional components should be. But if a model of the extended theory is given, it is easy to restrict it to (or: take a *reduct* consisting of) those components covered by the original theory.

Definition 2.1.2 A specification frame *representation* $\mu: \mathcal{F} \longrightarrow \mathcal{F}'$ consists of

- a functor $\Phi: \mathbf{Th} \longrightarrow \mathbf{Th}'$ and
- a natural transformation $\beta: \mathbf{Mod}' \circ \Phi^{op} \longrightarrow \mathbf{Mod}$

Composition of representations $\mathcal{F} \xrightarrow{\mu''} \mathcal{F}'' = \mathcal{F} \xrightarrow{\mu} \mathcal{F}' \xrightarrow{\mu'} \mathcal{F}''$ is defined by $\Phi'' = \Phi' \circ \Phi$ and $\beta'' = \beta \circ (\beta'_{\Phi^{op}})$.[2] This gives us a (quasi-)category **SpecFram** of specification frames. □

2.2 Institutions

Institutions introduced by Goguen and Burstall [GB92] split theories into signatures and sentences, thus the area of logic starts here:

Definition 2.2.1 An *institution* $\mathcal{I} = (\mathbf{Sign}, sen, \mathbf{Mod}, \models)$ consists of

- a category of *signatures* **Sign**,
- a functor $sen: \mathbf{Sign} \longrightarrow \mathbf{Set}$ giving the set of *sentences* over a given signature,

---

[1]**CAT** is the quasicategory of all categories. Quasicategories are categories that are allowed to have more than a class of objects, see Section A.5.

[2]Composition of morphisms is written in applicative order. $\beta'_{\Phi^{op}}$ is the natural transformation with $(\beta'_{\Phi^{op}})_\Sigma = \beta'_{\Phi^{op}(\Sigma)} = \beta'_{\Phi(\Sigma)}$.

2.2. INSTITUTIONS

- a functor $\mathbf{Mod}\colon (\mathbf{Sign})^{op} \longrightarrow \mathbf{CAT}$ giving the category of *models* of a given signature,

- for each $\Sigma \in |\mathbf{Sign}|$, a satisfaction relation $\models_\Sigma \subseteq |\mathbf{Mod}(\Sigma)| \times sen(\Sigma)$

such that for each morphism $\sigma\colon \Sigma \longrightarrow \Sigma'$ in $\mathbf{Sign}$ the *satisfaction condition*

$$M' \models_{\Sigma'} \sigma(\varphi) \iff M'|_\sigma \models_\Sigma \varphi$$

holds for each model $M' \in |\mathbf{Mod}(\Sigma')|$ and each sentence $\varphi \in sen(\Sigma)$, where $\sigma(\varphi)$ stands for $sen(\sigma)(\varphi)$. We call the "$\Longrightarrow$"-half of the satisfaction condition the principle that *reduction preserves satisfaction* (rps), and the "$\Longleftarrow$"-half the principle that *expansion preserves satisfaction* (eps). An *rps pre-institution* is given by the same data as an institution, but only rps is required.[3] □

Semantical entailment is defined as usual: For $\varphi \in sen(\Sigma)$, $\Gamma \subseteq sen(\Sigma)$, we write $\Gamma \models_\Sigma \varphi$ if all models that satisfy Γ also satisfy φ.

Definition 2.2.2 Given institutions $\mathcal{I} = (\mathbf{Sign}, sen, \mathbf{Mod}, \models)$ and $\mathcal{I}' = (\mathbf{Sign}', sen', \mathbf{Mod}', \models')$, a (plain) institution *representation* [Tar96a] (*plain map of institutions* in [Mes89]) $\mu = (\Phi, \alpha, \beta)\colon \mathcal{I} \longrightarrow \mathcal{I}'$ consists of

- a functor $\Phi\colon \mathbf{Sign} \longrightarrow \mathbf{Sign}'$,

- a natural transformation $\alpha\colon sen \longrightarrow sen' \circ \Phi$ and

- a natural transformation $\beta\colon \mathbf{Mod}' \circ \Phi^{op} \longrightarrow \mathbf{Mod}$

such that the following *representation condition* is satisfied for $M' \in |\mathbf{Mod}'(\Phi(\Sigma))|$ and $\varphi \in sen(\Sigma)$:

$$M' \models'_{\Phi(\Sigma)} \alpha_\Sigma(\varphi) \iff \beta_\Sigma(M') \models_\Sigma \varphi$$

Again, we can split this representation condition into an rps- and an eps-half. For representation between rps pre-institutions, only the rps-half is required. Let $\mu = (\Phi, \alpha, \beta)\colon \mathcal{I} \longrightarrow \mathcal{I}'$ and $\mu' = (\Phi', \alpha', \beta')\colon \mathcal{I}' \longrightarrow \mathcal{I}''$ be two institution representations. Then the composition $\mu'' = \mu' \circ \mu\colon \mathcal{I} \longrightarrow \mathcal{I}''$ consists of the following components:

- $\Phi'' = \Phi' \circ \Phi$

---

[3]The terminology (pre-institutions, rps) is taken from Scollo [SS92], with the slight modification that Scollo's pre-institutions have model functors going to $\mathbf{Set}$, while I keep them going to $\mathbf{CAT}$.

- $\alpha''_\Sigma = \alpha'_{\Phi\Sigma} \circ \alpha_\Sigma$ ($\Sigma \in |\mathbf{Sign}|$)

- $\beta''_\Sigma = \beta_\Sigma \circ \beta'_{\Phi\Sigma}$ ($\Sigma \in |\mathbf{Sign}|$)

This gives us a (quasi-)category **PlainInst** of institutions and (plain) representations and a category **PlainInstrps** of rps pre-institutions and (plain) representations. (We call representations *plain* in order to distinguish them from more complex forms of representations introduced in Chapter 4). □

An example of an rps pre-institution that is not an institution is many-sorted equational logic with behavioural satisfaction. Ehrig et al. [EBCO92] treat this logical system as a specification frame, but the proof of being a specification frame goes along the lines of the proof that **Th$_0$** is a functor from **PlainInstrps** to **SpecFram** below.

Definition 2.2.3 We define a functor $frame$: **PlainInstrps** $\longrightarrow$ **SpecFram** going from rps pre-institutions to specification frames which takes as theories of the specification frame just the signatures of the pre-institution: Let $frame(\mathbf{Sign}, sen, \mathbf{Mod}, \models)$ be just $(\mathbf{Sign}, \mathbf{Mod})$ (in general, a very poor specification frame), and $frame(\Phi, \alpha, \beta) = (\Phi, \beta)$. □

Definition 2.2.4 Another functor is a bit richer: it really combines signatures and sentences to get theories. An rps pre-institution $\mathcal{I} = (\mathbf{Sign}, sen, \mathbf{Mod}, \models)$ induces the category of *theories with axiom-preserving theory morphisms* $Th_0(\mathcal{I})$.
Objects are theories $T = (\Sigma, \Gamma)$, where $\Sigma \in |\mathbf{Sign}|$ and $\Gamma \subseteq sen(\Sigma)$ (with Γ not necessarily closed under consequence). Morphisms $\sigma: (\Sigma, \Gamma) \longrightarrow (\Sigma', \Gamma')$ in $Th_0(\mathcal{I})$ are signature morphisms $\sigma: \Sigma \longrightarrow \Sigma'$ such that $\sigma[\Gamma] \subseteq \Gamma'$, that is, axioms are mapped to axioms.

We can extract the components of a theory by setting $ax(\Sigma, \Gamma) = \Gamma$ and define a forgetful functor $sign: Th_0(\mathcal{I}) \longrightarrow \mathbf{Sign}$ which simply projects to the first component. We extend the functor $\mathbf{Mod}: \mathbf{Sign}^{op} \longrightarrow \mathbf{CAT}$ to $\mathbf{Mod}^{\mathbf{Th_0}}: Th_0(\mathcal{I})^{op} \longrightarrow \mathbf{CAT}$, where $\mathbf{Mod}^{\mathbf{Th_0}}(\Sigma, \Gamma)$ is the full subcategory of $\mathbf{Mod}(\Sigma)$ consisting of those M with $M \models_\Sigma \Gamma$ and for $\sigma: (\Sigma, \Gamma) \longrightarrow (\Sigma', \Gamma')$ in $Th_0(\mathcal{I})$, $\mathbf{Mod}^{\mathbf{Th_0}}(\sigma)$ is the restriction of $\mathbf{Mod}(\sigma)$ to $\mathbf{Mod}^{\mathbf{Th_0}}(\Sigma', \Gamma')$. We have to show that this restriction of $\mathbf{Mod}(\sigma)$ really ends in $\mathbf{Mod}^{\mathbf{Th_0}}(\Sigma, \Gamma)$. For $M' \in \mathbf{Mod}^{\mathbf{Th_0}}(\Sigma', \Gamma')$, that is, $M \models_{\Sigma'} \Gamma'$, since $\sigma[\Gamma] \subseteq \Gamma'$, we have $M \models_{\Sigma'} \sigma[\Gamma]$, and from this we get by the rps-half of the satisfaction condition (extended to sets of sentences) $M'|_\sigma \models_\Sigma \Gamma$, that is, $M'|_\sigma \in \mathbf{Mod}^{\mathbf{Th_0}}(\Sigma, \Gamma)$. So we end up with a specification frame $\mathbf{Th_0}(\mathcal{I}) = (Th_0(\mathcal{I}), \mathbf{Mod}^{\mathbf{Th_0}})$.

Now **Th₀** can be extended to representations: Given an rps pre-institution representation $\mu = (\Phi, \alpha, \beta): \mathcal{I} \longrightarrow \mathcal{I}'$, let Φ^α, the α-extension to theories of Φ, map (Σ, Γ) to $(\Phi(\Sigma), \alpha_\Sigma[\Gamma])$. By a similar argument as above, using the rps-half of the representation condition, β_Σ can easily be restricted to model categories of theories. This restriction is denoted by $\beta_{(\Sigma, \Gamma)}^{\mathbf{Th_0}}$. By putting $\mathbf{Th_0}(\mu) = (\Phi^\alpha, \beta^{\mathbf{Th_0}})$, **Th₀**: **PlainInstrps** $\longrightarrow$ **SpecFram** becomes a functor. □

By composing with the obvious inclusion of **PlainInst** into **PlainInstrps**, we get two functors from **PlainInst** to **SpecFram**, which also are denoted by *frame* and **Th₀**, respectively.

We will also need the *presentations functor* **Pres**: **PlainInstrps** $\longrightarrow$ **SpecFram**, which is defined like **Th₀**, except that we here restrict ourselves to theories with a *finite* set of axioms (called presentations).

2.3 Composable signatures and amalgamation

> "Given a species of structure, say widgets, then the result of interconnecting a system of widgets to form a super-widget corresponds to taking the *colimit* of the diagram of widgets in which the morphisms show how they are interconnected." *J. Goguen* [Gog91]

If we want to apply this to combine signatures and theories, we have to assume that signature categories are cocomplete. Moreover, this combination should be reflected by the semantics, that is, the model functor should be continuous. But there are very common institutions such as unsorted logics or type theories, which do not have a name for all semantical objects (for example, in unsorted logic, the carrier set is left implicit). This generally destroys continuity of the model functor. But the model functors of these institutions often do preserve a subclass of colimit, namely multiple pushouts (see A.2). Since in most cases, multiple pushouts suffice for the combination of signatures and theories, we here only talk about them. But most results easily generalize to all colimits (note that all colimits can be generated by multiple pushouts and initial objects).

Definition 2.3.1 A logical system is said to have *(finitely) composable signatures* (or theories, if we talk about specification frames), iff its signature (or theory) category has (finite) canonical multiple pushouts. A logical system representation is said to *preserve (finitely) composable signatures* (resp. theories), iff its Φ-component preserves (finite) multiple pushouts.

Given any category $\mathcal{C}$ of logical systems defined above, let $\mathcal{C}^{comp}$ denote the subcategory consisting of those objects with finitely composable theories and those representations preserving them. □

Definition 2.3.2 A specification frame, institution or rps pre-institution with finitely composable theories (or signatures or signatures, resp.) *has amalgamation*, if the model functor **Mod** preserves finite multiple pullbacks, that is, finite multiple pushouts in **Th** (or **Sign** or **Sign**, resp.) are taken to finite multiple pullbacks in **CAT**. Note that a multiple pullback category in **CAT** has as objects and morphisms amalgamations of objects and morphisms of the component categories, see A.2. □

It is not so clear what preservation of amalgamation by representations should mean. On the syntactical side, representations have to preserve multiple pushouts of signatures. What is the corresponding condition on the semantical side?

A straightforward answer is that amalgamations shall be possible not only along reduct functors, but also along mixtures of reducts and model translations. Technically, the requirement is the following: For a theory morphism $\sigma\colon T \longrightarrow T'$, diagrams

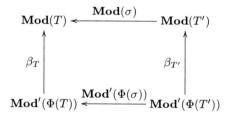

shall not only commute (which is guaranteed by naturality of β), but also have to be pullbacks in **CAT**.

We say that (Φ, β) *preserves amalgamation*, if, given a theory morphism $\sigma\colon T \longrightarrow T'$, the above diagram is a pullback (see A.2). For objects, this means that for any $M_1 \in \mathbf{Mod}'(\Phi(T))$ and $M_2 \in \mathbf{Mod}(T')$ with $\beta_T(M_1) = \mathbf{Mod}(\sigma)(M_2)$, there is a unique model $M_3 \in \mathbf{Mod}'(\Phi(T'))$ with $\mathbf{Mod}'(\Phi(\sigma))(M_3) = M_1$ and $\beta_{T'}(M_3) = M_2$.

Let **SpecFram**^{amal} be the restriction of **SpecFram** to those specification frames having *amalgamation* and those representations preserving amalgamation. Let **PlainInstrps**^{amal} be the restriction of **PlainInstrps** to those rps pre-institutions that have *amalgamation* and those representations preserving amalgamation (where this means that the above condition holds for *signature* morphisms). Similarly for **PlainInst**^{amal}.

2.3. COMPOSABLE SIGNATURES AND AMALGAMATION

Proposition 2.3.3 $\mathbf{Th_0}\colon \mathbf{PlainInstrps} \longrightarrow \mathbf{SpecFram}$ can be restricted to

$$\mathbf{Th_0}\colon \mathbf{PlainInstrps}^{comp} \longrightarrow \mathbf{SpecFram}^{comp}$$

(and similarly for **Pres**).

Proof:
This is the well-known fact that composable signatures induce composable theories, see [GB92]. Let $\mathcal{I} \in \mathbf{PlainInstrps}$ and $\mathcal{S} = ((\Sigma, \Gamma), ((\Sigma, \Gamma) \xrightarrow{\sigma_i} (\Sigma_i, \Gamma_i))_{i \in I})$ be a finite source of theories in $\mathbf{Th_0}(\mathcal{I})$. Let $(\Sigma', \Sigma \xrightarrow{\rho} \Sigma', (\Sigma_i \xrightarrow{\rho_i} \Sigma')_{i \in I})$ be the colimit of $sign(\mathcal{S})$ in the signature category of $\mathcal{I}$. Then $Colim(\mathcal{S}) = \mathcal{C}$, where $\mathcal{C}$ is the cone with tip

$$(\Sigma', \rho[\Gamma] \cup \bigcup_{i \in I} \rho_i[\Gamma_i])$$

and injections $\rho, (\rho_i)_{i \in I}$. Given another cone $\mathcal{T} = ((\Sigma'', \Gamma''), \Sigma \xrightarrow{\theta} \Sigma'', (\Sigma_i \xrightarrow{\theta_i} \Sigma'')_{i \in I})$ over $\mathcal{S}$, by the colimiting property in the signature category of $\mathcal{I}$, there is a unique signature morphism $\zeta \colon \Sigma' \longrightarrow \Sigma''$ with $\zeta \circ \mathcal{S} = \mathcal{T}$. Since for each $i \in I$, θ_i is a theory morphism, we have $\zeta[\rho_i[\Gamma_i]] = \theta_i[\Gamma_i] \subseteq \Gamma''$. Therefore, ζ is a theory morphism as well.

Now let $\mu = (\Phi, \alpha, \beta) \colon \mathcal{I} \longrightarrow \mathcal{I}'$ be an rps pre-institution representation preserving composable signatures. We have to show that $\mathbf{Th_0}(\mu) = (\Phi^\alpha, \beta^{\mathbf{Th_0}})$ preserves composable theories. Given a source $\mathcal{S}$ as above, we have

$\Phi^\alpha(tip(Colim(\mathcal{S}))) =$
$(\Phi(\Sigma'), \alpha_{\Sigma'}[\rho[\Gamma] \cup \bigcup_{i \in I} \rho_i[\Gamma_i]]) =$
$(tip(Colim(\Phi(sign(\mathcal{S})))), \Phi(\rho)[\alpha_\Sigma[\Gamma]] \cup \bigcup_{i \in I} [\Phi(\rho_i)(\alpha_{\Sigma_i}(\Gamma_i))]) =$
$tip(Colim(\Phi(\mathcal{S}))).$

Preservation of injections directly follows from μ preserving composable signatures. □

In order to carry this over to the semantical side (i. e. amalgamation), we show the following proposition which may be interesting in its own right:

Proposition 2.3.4 For an rps pre-institution $\mathcal{I}$ with composable signatures, the following are equivalent:

(1) $\mathcal{I}$ satisfies eps (that is, $\mathcal{I}$ is an institution) and $\mathcal{I}$ has amalgamation.

(2) $\mathbf{Th_0}(\mathcal{I})$ has amalgamation.

Proof:
(1) ⇒ (2): Let $((\Sigma,\Gamma),((\Sigma,\Gamma) \xrightarrow{\sigma_i} (\Sigma_i,\Gamma_i))_{i\in I})$ be a source in $\mathbf{Th_0}(\mathcal{I})$. By proposition 2.3.3, $\mathcal{S}$ has a multiple pushout

$$((\Sigma',\Gamma'),(\Sigma,\Gamma) \xrightarrow{\rho} (\Sigma',\Gamma'),((\Sigma_i,\Gamma_i) \xrightarrow{\rho_i} (\Sigma',\Gamma'))_{i\in I}).$$

Let $(M_i \in \mathbf{Mod}(\Sigma_i,\Gamma_i))_{i\in I}$ be a family of models with $M_i|_{\sigma_i} = M$ ($i \in I$) for some fixed $M \in \mathbf{Mod}(\Sigma,\Gamma)$. Then in $\mathcal{I}$, take the amalgamation $\underset{M}{+}(M_i)_I \in \mathbf{Mod}(\Sigma')$. We have to show that $\underset{M}{+}(M_i)_I \models_{\Sigma'} \Gamma' = \rho[\Gamma] \cup \bigcup_{i\in I} \rho_i[\Gamma_i]$. But this follows from $(\underset{M}{+}(M_i)_I)|_{\rho_i} = M_i \models_{\Sigma_i} \Gamma_i$ and $(\underset{M}{+}(M_i)_I)|_\rho = M \models_\Sigma \Gamma$ by the eps-half of the satisfaction condition.

(2) ⇒ (1): For a signature morphism $\sigma\colon \Sigma \longrightarrow \Sigma'$, $M' \in \mathbf{Mod}(\Sigma')$ and $\varphi \in sen(\Sigma)$, let $M'|_\sigma \models_\Sigma \varphi$. Consider the pushout of theories (where ι and ι' are obvious inclusions)

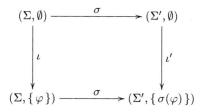

in $\mathbf{Th_0}(\mathcal{I})$. Then $M'|_\sigma \underset{M'|_\sigma}{+} M' \in \mathbf{Mod}(\Sigma', \{\sigma(\varphi)\})$. Since $M' = (M'|_\sigma \underset{M'|_\sigma}{+} M')|_{\iota'} = M'|_\sigma \underset{M'|_\sigma}{+} M'$, we have $M' \models_{\Sigma'} \sigma(\varphi)$.

That $\mathcal{I}$ has amalgamation follows from $\mathbf{Th_0}(\mathcal{I})$ having amalgamation. □

Proposition 2.3.5 Let $\mathcal{I}, \mathcal{I}'$ be rps pre-institutions which have amalgamation. For an rps pre-institution representation $\mu\colon \mathcal{I} \longrightarrow \mathcal{I}'$ preserving composable signatures, the following are equivalent:

(1) μ satisfies eps and preserves amalgamation.

(2) $\mathbf{Th_0}(\mu)$ preserves amalgamation.

Proof:
(1) ⇒ (2): Assume that an rps pre-institution representation $\mu =$

$(\Phi, \alpha, \beta): \mathcal{I} \longrightarrow \mathcal{I}'$ satisfies eps and preserves amalgamation, i. e. , for signature morphisms $\sigma: \Sigma \longrightarrow \Sigma'$, along the diagram

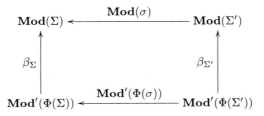

we have amalgamation. We have to show that for theory morphisms $\sigma: (\Sigma, \Gamma) \longrightarrow (\Sigma', \Gamma')$ in $\mathbf{Th_0}(\mathcal{I})$, the diagram

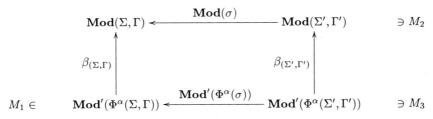

allows amalgamation as well. But this is easy: for $M_1 \in \mathbf{Mod}'(\Phi^\alpha(\Sigma, \Gamma))$ and $M_2 \in \mathbf{Mod}(\Sigma', \Gamma')$ with $\beta_{(\Sigma,\Gamma)}(M_1) = \mathbf{Mod}(\sigma)(M_2)$, by amalgamation along the signature morphism σ, there is a unique model $M_3 \in \mathbf{Mod}'(\Phi(\Sigma'))$ with $\mathbf{Mod}'(\Phi^\alpha(\sigma))(M_3) = \mathbf{Mod}'(\Phi(\sigma))(M_3) = M_1$ and $\beta_{(\Sigma',\Gamma')}(M_3) = \beta_{\Sigma'}(M_3) = M_2$. It remains to show that M_3 is a $\Phi^\alpha(\Sigma', \Gamma') = (\Phi(\Sigma'), \alpha_{\Sigma'}[\Gamma'])$-model. But $M_3 \models_{\Phi(\Sigma)} \alpha_{\Sigma'}[\Gamma']$ follows from $\beta_{\Sigma'}(M_3) = M_2 \models \Gamma'$ by the eps-half of the representation condition.

(2) $\Rightarrow$ (1): For $M' \in \mathbf{Mod}'(\Phi(\Sigma))$ and $\varphi \in sen(\Sigma)$, let $\beta_\Sigma(M') \models_\Sigma \varphi$. Now let ι be the inclusion of $(\Sigma, \emptyset)$ into $(\Sigma, \{\varphi\})$. Since $\mathbf{Th_0}(\mu)$ preserves amalgamation,

$$\begin{array}{ccc}
\mathbf{Mod}(\Sigma, \emptyset) & \xleftarrow{\mathbf{Mod}(\iota)} & \mathbf{Mod}(\Sigma, \{\varphi\}) \\
\beta_{(\Sigma,\emptyset)} \Big\uparrow & & \Big\uparrow \beta_{(\Sigma,\{\varphi\})} \\
\mathbf{Mod}'(\Phi(\Sigma), \emptyset) & \xleftarrow{\mathbf{Mod}'(\Phi(\iota))} & \mathbf{Mod}'(\Phi(\Sigma), \{\alpha_\Sigma(\varphi)\})
\end{array}$$

allows amalgamation. Therefore,

$$M' = M' \underset{\beta_\Sigma(M')}{+} \beta_\Sigma(M') \in \mathbf{Mod}'(\Phi(\Sigma), \{\alpha_\Sigma(\varphi)\}),$$

so $M' \models \alpha_\Sigma(\varphi)$.

That μ preserves amalgamation follows from $\mathbf{Th_0}(\mu)$ preserving amalgamation. □

Corollary 2.3.6 $\mathbf{Th_0}:\mathbf{PlainInst} \longrightarrow \mathbf{SpecFram}$ can be restricted to amalgamation: $\mathbf{Th_0}:\mathbf{PlainInst}^{amal} \longrightarrow \mathbf{SpecFram}^{amal}$. □

We next prove that representations with a model translation which is an isomorphism do preserve amalgamation.

Proposition 2.3.7 Let $\mu = (\Phi,\beta):\mathcal{F} \longrightarrow \mathcal{F}'$ be a specification frame representation or $\mu = (\Phi,\alpha,\beta):\mathcal{I} \longrightarrow \mathcal{I}'$ be an institution representation with Φ preserving multiple pushouts and β a natural isomorphism. Then μ preserves amalgamation.

Proof:
We consider the case for specification frames (for institutions $\mathcal{I}$, the result then follows by considering $frame(\mathcal{I})$).
Let β^{-1} be an inverse of β, and let $\sigma:T \longrightarrow T'$ be a theory morphism in $\mathcal{F}$.

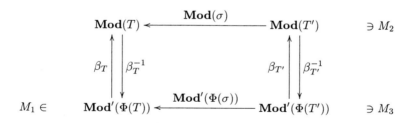

For $M_1 \in \mathbf{Mod}'(\Phi(T))$ and $M_2 \in \mathbf{Mod}(T')$ with $\beta_T(M_1) = \mathbf{Mod}(\sigma)(M_2)$, put $M_3 := \beta_{T'}^{-1}(M_2)$. Then $\beta_{T'}(M_3) = \beta_{T'}(\beta_{T'}^{-1}(M_2)) = M_2$, and $\mathbf{Mod}'(\Phi(\sigma))(M_3) = \beta_T^{-1}(\beta_T(\mathbf{Mod}'(\Phi(\sigma))(\beta_{T'}^{-1}(M_2)))) = \beta_T^{-1}(\mathbf{Mod}(\sigma)(\beta_{T'}(\beta_{T'}^{-1}(M_2)))) = \beta_T^{-1}(\mathbf{Mod}(\sigma)(M_2)) = \beta_T^{-1}(\beta_T(M_1)) = M_1$. Uniqueness of M_3 w. r. t. this property follows similarly. □

Propositions 2.3.4 and 2.3.5 as well as the results obtained in Chapter 4 suggest that the rps-half of the satisfaction condition is of a more syntactic nature, while the eps-half is a more semantical condition which interplays with amalgamation.

2.4 Entailment systems

Entailment systems, introduced by Meseguer [Mes89], are closely related to π-institutions [FS88]. They capture some aspects of proof theory.

Definition 2.4.1 An *entailment system* $\mathcal{E} = (\mathbf{Sign}, sen, \vdash)$ consists of

- a category **Sign** of *signatures*,
- a functor $sen\colon \mathbf{Sign} \longrightarrow \mathbf{Set}$ giving the set of *sentences* over a given signature, and
- for each $\Sigma \in |\mathbf{Sign}|$, an entailment relation $\vdash_\Sigma \subseteq \mathcal{P}(sen(\Sigma)) \times sen(\Sigma)$

such that the following properties are satisfied:

1. *reflexivity:* for any $\varphi \in sen(\Sigma)$, $\{\varphi\} \vdash_\Sigma \varphi$,
2. *monotonicity:* if $\Gamma \vdash_\Sigma \varphi$ and $\Gamma' \supseteq \Gamma$ then $\Gamma' \vdash_\Sigma \varphi$,
3. *transitivity:* if $\Gamma \vdash_\Sigma \varphi_i$, for $i \in I$, and $\Gamma \cup \{\varphi_i \mid i \in I\} \vdash_\Sigma \psi$, then $\Gamma \vdash_\Sigma \psi$,
4. $\vdash$-*translation:* if $\Gamma \vdash_\Sigma \varphi$, then for any $\sigma\colon \Sigma \longrightarrow \Sigma'$ in **Sign**, $\sigma[\Gamma] \vdash_{\Sigma'} \sigma(\varphi)$.

$\square$

Definition 2.4.2 Given entailment systems $\mathcal{E} = (\mathbf{Sign}, sen, \vdash)$ and $\mathcal{E}' = (\mathbf{Sign}', sen', \vdash')$, an entailment systems *representation* (called *plain map of entailment systems* by Meseguer) consists of

- a functor $\Phi\colon \mathbf{Sign} \longrightarrow \mathbf{Sign}'$ and
- a natural transformation $\alpha\colon sen \longrightarrow sen' \circ \Phi$

such that the following property is satisfied:

$$\Gamma \vdash_\Sigma \varphi \Rightarrow \alpha_\Sigma[\Gamma] \vdash'_{\Phi(\Sigma)} \alpha_\Sigma(\varphi)$$

The representation is called *conservative*, if in addition

$$\alpha_\Sigma[\Gamma] \vdash'_{\Phi(\Sigma)} \alpha_\Sigma(\varphi) \Rightarrow \Gamma \vdash_\Sigma \varphi$$

Composition of representations is like that for institution representations (only model translation is omitted). This gives a category **PlainEnt**. $\square$

2.5 Logics

Logics [Mes89] combine both model theory and proof theory:

Definition 2.5.1 A *logic* is a 5-tuple $\mathcal{L} = (\mathbf{Sign}, sen, \mathbf{Mod}, \vdash, \models)$ such that:

1. $(\mathbf{Sign}, sen, \vdash)$ is an entailment system (denoted by $ent(\mathcal{L})$),

2. $(\mathbf{Sign}, sen, \mathbf{Mod}, \models)$ is an institution (denoted by $inst(\mathcal{L})$), and

3. the following *soundness condition* is satisfied: for any $\Sigma \in |\mathbf{Sign}|$, $\Gamma \subseteq sen(\Sigma)$ and $\varphi \in sen(\Sigma)$,

$$\Gamma \vdash_\Sigma \varphi \Rightarrow \Gamma \models_\Sigma \varphi$$

A logic is *complete* if, in addition,

$$\Gamma \models_\Sigma \varphi \Rightarrow \Gamma \vdash_\Sigma \varphi$$

A logic representation $(\Phi, \alpha, \beta): \mathcal{L} \longrightarrow \mathcal{L}'$ is just an institution representation $(\Phi, \alpha, \beta) : inst(\mathcal{L}) \longrightarrow inst(\mathcal{L}')$ which is also an entailment system representation $(\Phi, \alpha): ent(\mathcal{L}) \longrightarrow ent(\mathcal{L}')$.

With composition taken from institution representations, this gives a category **PlainLog** of logics and logic representations. □

ent and log are easily extended to functors $ent: \mathbf{PlainLog} \longrightarrow \mathbf{PlainEnt}$ and $inst: \mathbf{PlainLog} \longrightarrow \mathbf{PlainInst}$. But there is also a functor $(\_)^+: \mathbf{PlainInst} \longrightarrow \mathbf{PlainLog}$ which makes a logic out of an institution. It is defined by $(\mathcal{I})^+ = (\mathbf{Sign}, sen, \mathbf{Mod}, \models, \models)$, where the second $\models$ is not satisfaction, but semantical entailment. $(\_)^+$ leaves representations unchanged. The representation $(Id_{\mathbf{Sign}}, Id_{sen}, Id_{\mathbf{Mod}}): \mathcal{L} \longrightarrow (inst(\mathcal{L})^+)$ is the unit of an adjunction between $inst$ and $(\_)^+$, see [CM93].

2.6 Transporting logical structure along maps

We state the fundamental definition of Cerioli and Meseguer [CM93]: Let $\underline{C}$ and $\underline{D}$ be categories and let $(\eta, \epsilon): U \dashv R: \underline{C} \longrightarrow \underline{D}$ be an adjoint situation. The idea is that $\underline{C}$ and $\underline{D}$ are categories of logical systems, $\underline{D}$ having richer structure and U forgetting this richer structure (though U is a left adjoint!). If for each morphism $c: C \longrightarrow U(D)$ in $\underline{C}$ the pullback

2.6. Transporting logical structure along maps

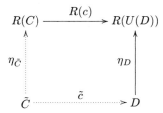

exists, we say that C *admits extension under R and U*, and $\tilde{c}\colon \tilde{C} \longrightarrow D$ is called the *extension of c by R and U*. Roughly speaking, $\tilde{C}$ is constructed out of C by enriching it with the features of D translated by c. Indeed, Cerioli and Meseguer show that this situation is given for all arrows in Fig. 1.1, except $frame\colon \textbf{Inst} \longrightarrow \textbf{SpecFram}$ (which is similar to the above introduced $frame\colon \textbf{PlainInst} \longrightarrow \textbf{SpecFram}$).

One application of this "borrowing" technique is the re-use of theorem provers. It is based on the following well-known theorem [AC92]:

Theorem 2.6.1 Let $\mu\colon \mathcal{I} \longrightarrow \mathcal{I}'$ be an institution representation with signature-wise surjective model translation components. Then semantical entailment is preserved:

$$\Gamma \models_\Sigma \varphi \text{ iff } \alpha_\Sigma[\Gamma] \models'_{\Phi(\Sigma)} \alpha_\Sigma(\varphi)$$

Proof:
$\Gamma \models_\Sigma \varphi$ iff for all $M \in \textbf{Mod}(\Sigma)$, $M \models_\Sigma \Gamma$ implies $M \models_\Sigma \varphi$ iff (by surjectivity) for all $M' \in \textbf{Mod}(\Sigma')$, $\beta_\Sigma(M') \models_\Sigma \Gamma$ implies $\beta_\Sigma(M') \models_\Sigma \varphi$ iff (by the representation condition) for all $M' \in \textbf{Mod}(\Sigma')$, $M' \models'_{\Phi(\Sigma)} \alpha_\Sigma[\Gamma]$ implies $M' \models'_{\Phi(\Sigma)} \alpha_\Sigma(\varphi)$ iff $\alpha_\Sigma[\Gamma] \models'_{\Phi(\Sigma)} \alpha_\Sigma(\varphi)$. □

So we can re-use a theorem-prover which is designed for $\mathcal{I}'$ also for proofs within $\mathcal{I}$. Cerioli and Meseguer [CM93] reformulate this in terms of institutions and logics: Given an institution representation $\mu = (\Phi, \alpha, \beta)\colon \mathcal{I} \longrightarrow inst(\mathcal{L}')$ into a logic $\mathcal{L}'$, take the following pullback in **PlainLog**:

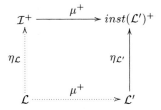

Then we have the following [CM93]:

Proposition 2.6.2   1. If $\mathcal{L}'$ is complete, then so is $\mathcal{L}$.

2. If β_Σ is surjective for each $\Sigma \in |\mathbf{Sign}|$, then $ent(\mu^+)$ is a conservative representation of the underlying entailment system, that is

$$\Gamma \vdash_\Sigma \varphi \Leftrightarrow \alpha_\Sigma[\Gamma] \vdash'_{\Phi(\Sigma)} \alpha_\Sigma(\varphi)$$

so we can re-use the entailment system of $\mathcal{L}'$ for $\mathcal{L}$

2.7 Liberality

The ACT ONE approach [EM85, EM90] to specification of parameterized abstract data types (PADTs) is as follows: A *parameterized theory* is a theory morphism $T \xrightarrow{\theta} T1$. A $T \xrightarrow{\theta} T1$-*parameterized abstract data type* is a pair (η, F), with $F: \mathbf{Mod}(T) \longrightarrow \mathbf{Mod}(T1)$ a functor and $\eta: Id_{\mathbf{Mod}(T)} \longrightarrow \mathbf{Mod}(\theta) \circ F$ a natural transformation. See [SST92] for a comparison to other approaches to parameterized abstract data types (which typically all use some sort of constraint, if not free generating constraints, then reachability, i. e. term-generating constraints).

Canonical PADTs are obtained by free constructions: A $T \xrightarrow{\theta} T1$-PADT (η, F) is called *free*, if F is left adjoint to $\mathbf{Mod}(\theta)$ with unit η, that is, for each $T1$-model B and each T-morphism $h: A \longrightarrow B|_\theta$, there exists a unique $T1$-morphism $h^\#: F(A) \longrightarrow B$ such that $h^\#|_\theta \circ \eta_A = h$.

Definition 2.7.1   An specification frame $\mathcal{F}$ is called *liberal*, if for each theory morphism $\sigma: T \longrightarrow T'$ in $\mathcal{F}$, $\mathbf{Mod}^{\mathbf{Th}_0(\mathcal{I})}(\sigma)$ has a left adjoint. An institution is called liberal, if the specification frame $\mathbf{Th}_0(\mathcal{I})$ is. □

Liberality means that free constructions that are needed for ACT ONE-style parameterization and module concepts [EM90] always exist. Moreover, we can use initiality and freeness constraints, which state that some part of a theory has to be interpreted in a fixed canonical way. The restriction to liberal institutions also has the effect that theorem proving becomes more feasible (conditional term rewriting, paramodulation), at least in the well-known liberal institutions.

Chapter 3

A variety of institutions of total, partial and order-sorted algebras

> "The need for a systematic treatment of partial operations is clear from practice. One must be able to handle *errors* and *exceptions*, and account for *non-terminating operations*. There are several approaches to deal with these in the literature, none of which appears to be fully satisfactory." *S. Feferman* [Fef92]

In this section, I formally introduce a variety of institutions. The emphasis is laid on institutions of many-sorted algebras that in some way capture the need of having operations that are only partially defined. The list of institutions is limited to liberal institutions, which have universally quantified conditional axioms as sentences, but contains the most expressive liberal institutions studied in the literature. (Note that there is a strong connection between liberality and axioms being universally quantified conditional, see [Tar85].)

3.1 Relational Partial Conditional Existence-Equational Logic

Let us first recall the institution $RP(R \stackrel{e}{=} \Rightarrow R \stackrel{e}{=})$ (Relational partial conditional existence-equational logic) of partial algebraic systems with universal Horn sentences over existential atomic formulas (see [Bur82, Bur86]).

This institution has a semantics of sentences that leads to a two-valued logic of partial algebras, see [Bur82]. Undefinedness of terms and falsehood of relations are not distinguished (as it would be possible within three-valued logics). An overview over different semantics and a justification of the two-valued semantics can be found in [Far91, CMR99].

A (finite) *signature* $\Sigma = (S, OP, POP, REL)$ consists of

- a (finite) set of sort symbols $s \in S$,
- a (finite) $S^* \times S$-indexed[1] set OP of total operation symbols,
- a (finite) $S^* \times S$-indexed set POP of partial operation symbols,
- a (finite) S^*-indexed set REL of relation, or predicate, symbols.

$w = s_1, \ldots, s_n$ is called the *arity*, s the *coarity* of an operation symbol $op \in OP_{w,s}$ (also written as $op\colon w \longrightarrow s$) or $pop \in POP_{w,s}$ (written as $pop\colon w \longrightarrow s$, and as $pop : w \longrightarrow ?s$ in specifications). Note that we have unrestricted overloading: for example, we may have $op\colon w \longrightarrow s \in OP$ and $op\colon w' \longrightarrow s' \in OP$ for $w \neq w'$, $s \neq s'$.

In the sequel, we will deal with finite signatures only, unless the contrary is stated explicitly (though many results generalize to the infinite case).

A *signature morphism* $\sigma\colon (S, OP, POP, REL) \longrightarrow (S', OP', POP', REL')$ consists of

- a map $\sigma^S\colon S \longrightarrow S'$,
- a map $\sigma_{w,s}^{OP}\colon OP_{w,s} \longrightarrow OP'_{\sigma^{S^*}(w), \sigma^S(s)}$[2] for each $w \in S^*, s \in S$,
- a map $\sigma_{w,s}^{POP}\colon POP_{w,s} \longrightarrow POP'_{\sigma^{S^*}(w), \sigma^S(s)}$ for each $w \in S^*, s \in S$, and
- a map $\sigma_w^{REL}\colon REL_w \longrightarrow REL'_{\sigma^{S^*}(w)}$ for each $w \in S^*$.

A Σ-*model* (or Σ-*algebra*) A consists of

- a family $|A| = (A_s)_{s \in S}$ of carrier sets,
- a family $(op_A\colon A_w \longrightarrow A_s)_{op\colon w \longrightarrow s \in OP}$ of total operations[3],
- a family $(pop_A\colon \text{dom } pop_A \longrightarrow A_s)_{pop\colon w \longrightarrow s \in POP}$ of partial operations, where dom $pop_A \subseteq A_w$ is the domain of pop_A, and

---
[1] S^* is the set of finite strings over S
[2] σ^{S^*} is the extension of σ^S to strings
[3] For $w = s_1, \ldots, s_n$, we abbreviate $A_{s_1} \times \cdots \times A_{s_n}$ by A_w.

3.1. RELATIONAL PARTIAL CONDITIONAL EXISTENCE-EQUATIONAL LOGIC

- a family $(R_A \subseteq A_w)_{R:w \in REL}$ of relations.

A Σ-*homomorphism* $h\colon A \longrightarrow B$ is a family $h = (h_s\colon A_s \longrightarrow B_s)_{s \in S}$ of total functions, such that for any $op\colon w \longrightarrow s \in OP$ and any $(a_1, \ldots, a_n) \in A_w$ we have
$$h_s(op_A(a_1, \ldots, a_n)) = op_B(h_{s_1}(a_1), \ldots, h_{s_n}(a_n)),$$
for any $pop\colon w \longrightarrow s \in POP$, $(a_1, \ldots, a_n) \in A_w$ and $a \in A_s$ we have
$$(a_1, \ldots, a_n) \in \text{dom } pop_A \text{ and } pop_A(a_1, \ldots, a_n) = a$$
implies
$$(h_{s_1}(a_1), \ldots, h_{s_n}(a_n)) \in \text{dom } pop_B \text{ and } pop_B(h_{s_1}(a_1), \ldots, h_{s_n}(a_n)) = h_s(a)$$
and for any $R : w \in REL$ and $(a_1, \ldots, a_n) \in A_w$ we have
$$(a_1, \ldots, a_n) \in R_A \Rightarrow (h_{s_1}(a_1), \ldots, h_{s_n}(a_n)) \in R_B$$

If $\Sigma = (S, OP, POP, REL)$, $\sigma\colon \Sigma \longrightarrow \Sigma'$ is a signature morphism and A' is a Σ'-model, then $A'|_\sigma$ is the Σ-algebra A with

- $A_s := A'_{\sigma^S(s)} \quad (s \in S)$
- $op_A := (\sigma^{OP}_{w,s}(op))_{A'} \quad (op\colon w \longrightarrow s \in OP)$
- $pop_A := (\sigma^{POP}_{w,s}(pop))_{A'} \quad (pop\colon w \longrightarrow s \in POP)$
- $R_A := (\sigma^{REL}_w(R))_{A'} \quad (R : w \in REL)$

Similarly, for homomorphisms we define $(h'|_\sigma)_s := h_{\sigma(s)}$.

A *variable system* X over a signature $\Sigma = (S, OP, POP, REL)$ is an S-sorted set $(X_s)_{s \in S}$ with the X_s pairwise disjoint. Given a variable system X, the set $T_\Sigma(X)_s$ of $\Sigma(X)$-*terms of sort* s is inductively defined:

1. $x \in T_\Sigma(X)_s$ for $x \in X_s$
2. "$op(t_1, \ldots, t_n) : s$" $\in T_\Sigma(X)_s$ for $op\colon s_1, \ldots, s_n \longrightarrow s \in OP$ and $t_i \in T_\Sigma(X)_{s_i}$, $i = 1, \ldots, n$.
3. "$pop(t_1, \ldots, t_n) : s$" $\in T_\Sigma(X)_s$ for $pop\colon s_1, \ldots, s_n \longrightarrow s \in POP$ and $t_i \in T_\Sigma(X)_{s_i}$, $i = 1, \ldots, n$.

If the third clause is omitted, we get the set of all *total* terms over Σ and X, denoted by $T^{tot}_\Sigma(X)$.

Note that unrestricted overloading makes the parsing of terms ambiguous, if no explicit sort annotations are given. But in cases when we do not use overloading (or just restricted forms of overloading), we can omit the sort annotations form terms.

Now $T_\Sigma(X)$ can be equipped with an algebraic structure by putting

- $op_{T_\Sigma(X)}(t_1, \ldots, t_n) := op(t_1, \ldots, t_n) \quad (op\colon s_1, \ldots, s_n \longrightarrow s \in OP)$
- $pop_{T_\Sigma(X)}(t_1, \ldots, t_n) := pop(t_1, \ldots, t_n) \quad (pop\colon s_1, \ldots, s_n \longrightarrow s \in POP)$
- $R_{T_\Sigma(X)} := \emptyset \quad (R \in REL)$

Then $T_\Sigma(X)$ becomes the *term algebra over Σ with variables X*.

A *valuation* $\nu\colon X \longrightarrow A$ is an S-map from a variable system X to the S-sorted set of carriers of a Σ-algebra A. Valuations have *homomorphic extensions* $\nu^\#\colon \operatorname{dom} \nu^\# \longrightarrow A$ to the set $\operatorname{dom} \nu^\# \subseteq T_\Sigma(X)$ of ν-*interpretable terms*:

1. Each variable $x \in X_s$, is ν-interpretable, and $\nu_s^\#(x) := \nu_s(x)$.

2. $op(t_1, \ldots, t_n)$ with $op\colon s_1, \ldots, s_n \longrightarrow s \in OP$ and $t_i \in T_\Sigma(X)_{s_i}$, $i = 1, \ldots, n$ is ν-interpretable, if for $i = 1, \ldots, n$, $t_i \in \operatorname{dom} \nu_{s_i}^\#$. In this case, we define $\nu_s^\#(op(t_1, \ldots, t_n)) := op_A(\nu_{s_1}^\#(t_1), \ldots, \nu_{s_1}^\#(t_1))$.

3. $pop(t_1, \ldots, t_n)$ with $pop\colon s_1, \ldots, s_n \longrightarrow s \in POP$ and $t_i \in T_\Sigma(X)_{s_i}$, $i = 1, \ldots, n$ is ν-interpretable, if

    (a) for $i = 1, \ldots, n$, $t_i \in \operatorname{dom} \nu_{s_i}^\#$ (all t_i are ν-interpretable) and
    (b) $(\nu_{s_1}^\#(t_1), \ldots, \nu_{s_1}^\#(t_1)) \in \operatorname{dom} pop_A$

    In this case, we define $\nu_s^\#(pop(t_1, \ldots, t_n)) := pop_A(\nu_{s_1}^\#(t_1), \ldots, \nu_{s_1}^\#(t_1))$.

A *sentence* over a signature $\Sigma = (S, OP, POP, REL)$ is a conditional formula

$$\forall X \,.\, e_1 \wedge \cdots \wedge e_n \Longrightarrow e$$

where X is a variable system over Σ and the atomic formulas e and e_i are of two kinds: Either

$$t_1 \stackrel{e}{=} t_2$$

where t_1, t_2 are terms with same sort from $T_\Sigma(X)_s$. Or

$$R(t_1, \ldots, t_n)$$

where $R: w \in REL$ and for $i = 1, \ldots, n$, t_i is a term of sort s_i (from $T_\Sigma(X)_{s_i}$).

3.1. RELATIONAL PARTIAL CONDITIONAL EXISTENCE-EQUATIONAL LOGIC

Sentence translation along signature morphisms $\sigma: \Sigma \longrightarrow \Sigma'$ is defined in the following way: Given a Σ-sentence

$$\varphi = \forall X \,.\, e_1 \wedge \cdots \wedge e_n \Longrightarrow e$$

translate X to the variable system $\sigma(X)$ with $\sigma(X)_{s'} = \bigcup_{\sigma(s)=s'} X_s$. (Here, the disjointness of the X_s is needed.) Define a valuation $\xi: X \longrightarrow T_{\Sigma'}(\sigma(X))|_\sigma$ by putting $\xi(x) = x$. Then $\xi^\#: T_\Sigma(X) \longrightarrow T_{\Sigma'}(\sigma(X))|_\sigma$ can be easily extended to atomic formulas by letting it act on the component terms. Now $\sigma(\varphi)$ is given by

$$\forall \sigma(X) \,.\, \xi^\#(e_1) \wedge \cdots \wedge \xi^\#(e_n) \Longrightarrow \xi^\#(e)$$

Finally, *satisfaction* is defined as follows: $A \models_\Sigma (X \,.\, e_1 \wedge \ldots \wedge e_n \Longrightarrow e)$ if and only if all valuations $\nu: X \longrightarrow A$ which satisfy the premises satisfy the conclusion as well. Satisfaction of atomic formulas is defined as

$$\nu \models t_1 \stackrel{e}{=} t_2 \iff \nu^\#(t_1) \text{ and } \nu^\#(t_2) \text{ are both defined and equal}$$

$$\nu \models R(t_1, \ldots, t_n) \iff (\nu^\#(t_1), \ldots, \nu^\#(t_n)) \text{ is defined and } \in R_A$$

Since X and $\sigma(X)$ contain exactly the same variables (only under different sorts), given a signature morphism $\sigma: \Sigma \longrightarrow \Sigma'$ and a Σ'-model A', there is a one-one-correspondence between valuations $\nu: X \longrightarrow A'|_\sigma$ and $\nu': \sigma(X) \longrightarrow A'$ that makes

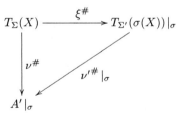

commute in the category of S-sorted sets and S-sorted partial functions (that is, $\nu^\#(t)$ is defined iff $\nu'^\#(\xi^\#(t))$ is defined, and in that case, $\nu^\#(t) = \nu'^\#(\xi^\#(t))$). From this, the satisfaction condition follows easily.

The proof of the satisfaction condition may be simplified by presenting the institution as a parchment (see chapter 8). □

For sets of variables, we use the following notation: $X = \{\, x : s_1; y : s_2 \,\}$ means that $X_{s_1} = \{\, x \,\}$, $X_{s_2} = \{\, y \,\}$ and $X_s = \emptyset$ for $s \neq s_1$ and $s \neq s_2$. Valuations $\nu: \{\, x_1 : s_1, \ldots, x_n : s_n \,\} \longrightarrow A$ will sometimes considered to be arguments for operations $op: s_1 \times \cdots \times s_n \longrightarrow s$ in A.

3.2 Restrictions of $RP(R \stackrel{e}{=}\!\!\Rightarrow R \stackrel{e}{=})$

Some natural restrictions of $RP(R \stackrel{e}{=}\!\!\Rightarrow R \stackrel{e}{=})$ can now be defined immediately:

$P(\stackrel{e}{=}\!\!\Rightarrow\stackrel{e}{=})$ (Partial Conditional Existence Equational Logic) is the subinstitution of $RP(R \stackrel{e}{=}\!\!\Rightarrow R \stackrel{e}{=})$ defined by requiring $REL = \emptyset$ for signatures $\Sigma = (S, OP, POP, REL)$. □

Let $P(D \Rightarrow\stackrel{e}{=})$ (Partial Existentially-Conditioned Existence-Equational Logic) be the subinstitution of $P(\stackrel{e}{=}\!\!\Rightarrow\stackrel{e}{=})$ defined by requiring sentences to have the form

$$\forall X \,.\, t_1 \stackrel{e}{=} t_1 \wedge \cdots \wedge t_n \stackrel{e}{=} t_n \Longrightarrow t \stackrel{e}{=} t'$$

that is, the premises contain only definedness conditions. $P(D \Rightarrow\stackrel{e}{=})$ was introduced by Burmeister [Bur82]. But also Jarzembski's weak varieties [Jar88, Jar93], when restricted to the liberal case, are of this form. □

$P(\stackrel{s}{=})$ is the institution with signatures and models from $P(\stackrel{e}{=}\!\!\Rightarrow\stackrel{e}{=})$. A Σ-sentence has the form

$$\forall X \,.\, t_1 \stackrel{s}{=} t_2 \quad (t_1, t_2 \in T_\Sigma(X)_s)$$

Satisfaction is defined as follows: $A \models_\Sigma \forall X \,.\, t_1 \stackrel{s}{=} t_2$ iff for all valuations $\nu \colon X \longrightarrow A$, $\nu^\#(t_1)$ is defined if and only if $\nu^\#(t_2)$ is defined, and if this is the case, both are equal.

These "strong equations" are cited in the literature quite frequently, see [Bur82, Hoe81, Kle52, Slo68]. □

Note that in the literature, total operations are often omitted in the presence of partial operations: a total operation can be viewed as a partial operation satisfying $pop(x_1, \ldots, x_n) \stackrel{e}{=} pop(x_1, \ldots, x_n)$, if existence equations are available. In $P(\stackrel{s}{=})$, there are no existence equations, and in fact, $P(\stackrel{s}{=})$ without total operations is slightly weaker than with: the algebra with totally undefined partial operations is a model of every sentence containing operation symbols on both sides, while this does not happen in the presence of a total operation op with the same arity and coarity as a partial operation pop: then the equation

$$\forall x \colon s \,.\, op(x) \stackrel{s}{=} pop(x)$$

forces pop to be total, so $\forall x \colon s \,.\, pop(x) \stackrel{e}{=} pop(x)$ holds as well.

$HEP(\stackrel{e}{=}\!\!\Rightarrow\stackrel{e}{=})$ (for Hierarchical Equationally Partial Theories) defined by Reichel [Rei87] is the institution with signatures $\Sigma = (S, OP, POP, \preceq, \mathit{Def})$ consisting of

$P(\overset{e}{=}\Rightarrow\overset{e}{=})$-signatures (S, OP, POP) together with a well-founded partial order $\preceq$ on POP and a mapping Def, which assigns to each $pop: s_1, \ldots, s_n \longrightarrow s \in POP$ a finite set of existence equations

$$Def(pop) = X.E \text{ with } E \subseteq T_\Sigma(X) \times T_\Sigma(X)$$

where $X = \{x_1 : s_1, \ldots, x_n : s_n\}$, and such that all partial operation symbols occurring in $Def(pop)$ are strictly less than pop w. r. t. $\preceq$.

Signature morphisms are those from $P(\overset{e}{=}\Rightarrow\overset{e}{=})$ with the additional requirement that

$$Def(\sigma(pop)) = \sigma \times \sigma(Def(pop))$$

Σ-models are (S, OP, POP)-models A in $P(\overset{e}{=}\Rightarrow\overset{e}{=})$ such that dom pop_A equals

$\{(\nu(x_1), \ldots, \nu(x_n)) \in A_w \mid$
$\quad \nu: X \longrightarrow A$ with $\nu \models t_1 \overset{e}{=} t_2$ for all $X.t_1 \overset{e}{=} t_2 \in Def(pop)\}$

Reducts, sentences and satisfaction are defined as in $P(\overset{e}{=}\Rightarrow\overset{e}{=})$.

Note that Reichel's theory morphisms defined in [Rei87] are slightly more general than theory morphisms in $HEP(\overset{e}{=}\Rightarrow\overset{e}{=})$. But Reichel only defines a specification frame. If signatures and sentences have to be separated, a slight restriction has to be made.

Freyd's essentially algebraic theories [Fre72] are essentially the same (Freyd does not define signatures, reducts and the like).

We define two restrictions of $HEP(\overset{e}{=}\Rightarrow\overset{e}{=})$: First, let $HEP1(\overset{e}{=}\Rightarrow\overset{w}{=})$ be the restriction to those signatures where the hierarchy of operations has only height one, that is, the domain of each partial operation is defined using total operations only. Axioms have to be of form

$$\forall X . \; t_1 \overset{e}{=} u_1 \wedge \cdots \wedge t_n \overset{e}{=} u_n \Longrightarrow t \overset{w}{=} u$$

where $t \overset{w}{=} u$ is a weak equation, which is satisfied if in case of definedness of both sides, they are equal. $HEP1(\overset{w}{=})$ is the restriction of $HEP1(\overset{e}{=}\Rightarrow\overset{w}{=})$ to unconditional weak equations, and $HEP1(\Rightarrow\overset{w}{=})$ is the restriction of $HEP1(\overset{e}{=}\Rightarrow\overset{w}{=})$ to conditional equations with only total terms in the premises. □

3.3 Further restrictions of $RP(R \overset{e}{=}\Rightarrow R \overset{e}{=})$

We now further restrict $RP(R \overset{e}{=}\Rightarrow R \overset{e}{=})$ and get institutions which will turn out to be strictly weaker in expressiveness.

3. A VARIETY OF INSTITUTIONS OF TOTAL, PARTIAL AND ORDER-SORTED ALGEBRAS

institution	relations	partiality	atomic formulas				axioms
			total equations	flat partial eq.	partial equations	rel. applications	
$RP(R \stackrel{e}{=}\!\!\Rightarrow R \stackrel{e}{=})$	x	x	x	x	x	x	conditional
$P(\stackrel{e}{=}\!\!\Rightarrow\stackrel{e}{=})$	–	x	x	x	x	–	conditional
$RP(R \stackrel{f}{=}\!\!\Rightarrow R \stackrel{f}{=})$	x	x	x	x	–	x	conditional
$P(\stackrel{f}{=}\!\!\Rightarrow\stackrel{f}{=})$	–	x	x	x	–	–	conditional
$R(R \Rightarrow R =)$	x	–	x	–	–	x	conditional
$(=\!\!\Rightarrow=)$	–	–	x	–	–	–	conditional
$RP(R \stackrel{e}{=})$	x	x	x	x	x	x	unconditional
$P(\stackrel{e}{=})$	–	x	x	x	x	–	unconditional
$RP(R \stackrel{f}{=})$	x	x	x	x	–	x	unconditional
$P(\stackrel{f}{=})$	–	x	x	x	–	–	unconditional
$R(R=)$	x	–	x	–	–	x	unconditional
$(=)$	–	–	x	–	–	–	unconditional

Let $RP(=\!\!\Rightarrow=)(R \stackrel{f}{=})$ be $RP(R \stackrel{f}{=})$ with all sentences from $(=\!\!\Rightarrow=)$ added.

"Flat" partiality means that atomic formulas in axioms are of form

$$pop(t_1, \ldots, t_n) \stackrel{e}{=} t_0 \quad \text{or} \quad R(t_1, \ldots, t_n)$$

where $t_1, \ldots, t_n$ and t_0 consist of total operation symbols only. That is, there is no nested partiality and no partiality combined with relations. This is indicated with the $\stackrel{f}{=}$-sign. Note that in this case, we can give a definition of satisfaction just in terms of *total* homomorphic extensions of valuations, since partial operations symbols are applied only to total terms: A valuation $\nu: X \longrightarrow A$ satisfies $pop(t_1, \ldots, t_n) \stackrel{e}{=} t$ if and only if $(\nu^\#(t_1), \ldots, \nu^\#(t_n)) \in \mathrm{dom}\, pop_A$ and $pop_A(\nu^\#(t_1), \ldots, \nu^\#(t_n)) = \nu^\#(t)$. Similarly for relations.

The =-sign is used for the restriction to equations between total terms.

Concerning weak equations, let us introduce $P(\stackrel{e}{=}\stackrel{w}{=})$ which extends $P(\stackrel{e}{=})$ by also allowing unconditional weak equations, and $P(\stackrel{e}{=}\!\!\Rightarrow\stackrel{w}{=})(\stackrel{e}{=})$, which additionally allows weak equations conditioned by existence-equations.

Now $R(R \Rightarrow R =)$ (Horn Clause Logic) is the basis for the specification language Eqlog [GM86, Pad88]), its restriction $R(R \Rightarrow R)$ to axioms with relational conclusion is the basis of logic programming and Prolog [Llo87]. $(=\!\!\Rightarrow=)$ (Conditional Equational Logic) and $(=)$ (Equational Logic) are well-known institutions as well [GTW78, TWW81, EM85]. The other institutions will later turn out to be equivalent in expressiveness to well-known ones.

3.4 Some restrictions with special interpretation

Equational Type Logic ($\mathcal{ETL}$) due to Manca, Salibra and Scollo [MSS90] is a (plain) subinstitution of $R(R \Longrightarrow R =)$: An $\mathcal{ETL}$-signature is a one-sorted $R(R \Longrightarrow R =)$-signature with exactly one relation symbol : (the typing relation), which is binary. Models, reducts, sentences and satisfaction are inherited from $R(R \Longrightarrow R =)$.

Unified algebras ($\mathcal{UNIFALG}$) due to Mosses [Mos89] form a subinstitution of $\mathbf{Th_0}(R(R \Longrightarrow R =))$ (that is, a simple subinstitution of $R(R \Longrightarrow R =)$, in the sense of chapter 4). A unified algebra-signature is a one-sorted $R(R \Longrightarrow R =)$-theory $((S, OP, REL), \Gamma)$ where (S, OP, REL) contains the following signature

sorts u
ops $nothing : u$
$\_|\_: u \times u \longrightarrow u$
$\_\&\_: u \times u \longrightarrow u$
preds $\leq: u \times u$
$\_:\_: u \times u$

and Γ consists of exactly the following axioms:

$\forall x, y.\ x \leq y \wedge y \leq x \Longrightarrow x = y$
$\forall x, y, z.\ x \leq y \wedge y \leq z \Longrightarrow x \leq z$
$\forall x.\ x \leq x$
$\forall x.\ nothing \leq x$
$\forall x, y, z.\ x \leq z \wedge y \leq z \Longrightarrow x|y \leq z$
$\forall x, y.\ x \leq x|y$
$\forall x, y.\ y \leq x|y$
$\forall x, y, z.\ z \leq x \wedge z \leq y \Longrightarrow z \leq x\&y$
$\forall x, y, z.\ x\&(y|z) = (x\&y)|(x\&z)$
$\forall x, y, z.\ x|(y\&z) = (x|y)\&(x|z)$
$\forall x, y.\ x : x \wedge x \leq y \Longrightarrow x : y$
$\forall x, y.\ x : y \Longrightarrow x : x$
$\forall x, y.\ x : y \Longrightarrow x \leq y$
$\forall x_1, \ldots, x_n, x_i'.\ x_i \leq x_i' \Longrightarrow$
$\quad op(x_1, \ldots, x_i, \ldots, x_n) \leq op(x_1, \ldots, x_i', \ldots, x_n)$
$\quad \text{for } op: u^n \longrightarrow u \in OP, i = 1, \ldots, n$

Models, reducts[4], sentences and satisfaction are inherited from $\mathbf{Th_0}(R(R \Longrightarrow R =))$.

Of course, this captures only part of the intended meaning, since in $\mathcal{ETL}$ and

---

[4]Note that Mosses defined two kinds of reducts. Only his forgetful functors, but not his "more forgetful functors" can be formalized in a subinstitution of $\mathbf{Th_0}(R(R \Longrightarrow R =))$.

$\mathcal{UNIFALG}$, the typing relation : has a special meaning. The intended meaning can in some cases be formalized by an institution representation with codomain $\mathcal{UNIFALG}$ resp. $\mathcal{ETL}$, see the discussion of categorical retractive simulations in chapter 7.

3.5 Limit theories

$R(R \Longrightarrow \exists!R =)$ is the following extension of the institution $R(R \Longrightarrow R =)$: Signatures $\Sigma = (S, OP, REL)$ are $R(R \Longrightarrow R =)$-signatures. Also, signature morphisms, models and reducts are inherited from $R(R \Longrightarrow R =)$. Σ-sentences have the form

$$\forall X.\ e_1 \wedge \cdots \wedge e_n \Longrightarrow \exists!Y.\ (e'_1 \wedge \cdots \wedge e'_m)$$

where X and Y are disjoint S-sorted variable sets and the e_i are atomic formulas with terms over X, while the e'_i are atomic formulas with terms over $X \cup Y$. The atomic formulas may be either relation applications or equalities. Coste's limit theories [Cos79] have more general sentences, but can be shown to be equivalent to conjunctions of sentences in the above form.

Satisfaction is defined as follows:

$$A \models_\Sigma \forall X.\ e_1 \wedge \cdots \wedge e_n \Longrightarrow \exists!Y.\ (e'_1 \wedge \cdots \wedge e'_m)$$

if for all valuations $\nu\colon X \longrightarrow A$ satisfying all the e_i ($i = 1, \ldots, n$), there exists a unique extension $\xi\colon X \cup Y \longrightarrow A$ of ν which satisfies all the e'_i ($i = 1, \ldots, m$). □

$R(R \Longrightarrow =)(\exists!R =)$ is the restriction of $R(R \Longrightarrow \exists!R =)$ to those sentences which are either conditional sentences of form

$$\forall X.\ e_1 \wedge \cdots \wedge e_n \Longrightarrow t_1 = t_2$$

with the e_i being atomic formulas with terms over X and t_1, t_2 terms over X, or unconditional sentences of form

$$\forall X \exists!Y.\ (e'_1 \wedge \cdots \wedge e'_m)$$

where the e'_i are atomic formulas with terms over $X \cup Y$.

3.6 Some algebraic notions and propositions

The following algebraic notions apply to $RP(R \overset{e}{\Longrightarrow} R \overset{e}{=})$ and all its restrictions (though they may become trivial in the more restricted institutions: for example, in $(\Longrightarrow =)$, all homomorphisms are closed).

3.6. SOME ALGEBRAIC NOTIONS AND PROPOSITIONS

Given a Σ-algebra A and an S-subset $|A'|$ of $|A|$ that is closed under the total operations, then the *relative subalgebra* A' of A induced by $|A'|$, is defined as follows:

- $op_{A'}(a_1, \ldots, a_n) := op_A(a_1, \ldots, a_n)$ for $op\colon w \longrightarrow s \in OP$ and $(a_1, \ldots, a_n) \in A'_w$

- for $pop\colon w \longrightarrow s \in POP$, graph $(pop_{A'})$ = graph $(pop_A) \cap A'_w \times A'_s$, that is,
$$pop_{A'}(a_1, \ldots, a_n) := \begin{cases} pop_A(a_1, \ldots, a_n), & \text{if } pop_A(a_1, \ldots, a_n) \in A'_s \\ \text{undefined}, & \text{otherwise} \end{cases}$$

- for $R : w \in REL$, $R_{A'} := R_A \cap A'_w$

A *congruence* $\equiv$ on a Σ-algebra A is an S-relation $(\equiv_s \subseteq A_s \times A_s)_{s \in S}$ which is reflexive, symmetric, transitive and compatible with the operations. The latter means that for $op\colon w \longrightarrow s \in OP, (a_1, \ldots, a_n), (a'_1, \ldots, a'_n) \in A_w$ resp. for $op\colon w \longrightarrow s \in POP, (a_1, \ldots, a_n), (a'_1, \ldots, a'_n) \in \text{dom } op_A$ we have

If $((a_1, \ldots, a_n), (a'_1, \ldots, a'_n)) \in \equiv_w$,
then $(op_A(a_1, \ldots, a_n), op_A(a'_1, \ldots, a'_n)) \in \equiv_s$

The smallest congruence on A is $\Delta A = (\{(a,a) \mid a \in A_s\})_{s \in S}$, the diagonal (or identity) relation. The congruence *generated* by an S-relation $R \subseteq A \times A$, denoted by $<R>$, is the least congruence containing R, which always exists because the property of being a congruence is preserved by arbitrary intersections of relations. Given a Σ-homomorphism $h\colon A \longrightarrow B$, its *kernel* ker h defined by

$$(\ker h)_s := \{(a, a') \mid a, a' \in A_s, h_s(a) = h_s(a')\}$$

is a congruence.

A Σ-homomorphism $h\colon A \longrightarrow B$ is called *full*, if for all $pop\colon w \longrightarrow s \in POP$ and $(a_1, \ldots, a_n) \in A_w$

$(h_{s_1}(a_1), \ldots, h_{s_n}(a_n)) \in \text{dom } op_B$ and $op_B(h_{s_1}(a_1), \ldots, h_{s_n}(a_n)) \in h_s[A_s]$

implies

$\exists (a'_1, \ldots, a'_n) \in \text{dom } op_A$ with $h_{s_i}(a_i) = h_{s_i}(a'_i), i = 1, \ldots, n$

and for all $R : w \in REL$:

$(h_{s_1}(a_1), \ldots, h_{s_n}(a_n)) \in R_B$ implies $\exists (a'_1, \ldots, a'_n) \in R_A$
with $h_{s_i}(a_i) = h_{s_i}(a'_i), i = 1, \ldots, n$.

Stated succinctly:

$h_w \times h_s[graph(pop_A)] = graph(pop_B) \cap h_w \times h_s[A_w \times A_s]$ and $h_w[R_A] = R_B \cap h_w[A_w]$.

A Σ-homomorphism $h: A \longrightarrow B$ is called *closed*, if for all $pop: w \longrightarrow s \in POP$ and $(a_1, \ldots, a_n) \in A_w$

if $(h_{s_1}(a_1), \ldots, h_{s_n}(a_n)) \in \text{dom } op_B$, then $(a_1, \ldots, a_n) \in \text{dom } op_A$

and for all $R : w \in REL$:

if $(h_{s_1}(a_1), \ldots, h_{s_n}(a_n)) \in R_B$, then $(a_1, \ldots, a_n) \in R_A$.

Stated succinctly:

$$h_w^{-1}[\text{dom } op_B] = \text{dom } op_A \text{ and } h_w^{-1}[R_B] = R_A.$$

It can be easily shown that each closed homomorphism is also full.

Definition 3.6.1 Let A be Σ-algebra and $\equiv$ a congruence on A. Let $|A/\equiv|$ be the set-theoretical factorization of $|A|$ by $\equiv$, $[a]\equiv$ (or $[a]$, for short) the equivalence class of $a \in A_s$ and $\text{nat}_\equiv: |A| \longmapsto |A/\equiv|$ be the natural projection mapping a to $[a]\equiv$.

Now impose the following algebraic structure on $|A/\equiv|$ to get a Σ-algebra $A/\equiv$: For $R : s_1, \ldots, s_n \in REL$, $w = s_1, \ldots, s_n$ let

$$\begin{aligned} R_{A/\equiv} := \{ &([a_1], \ldots, [a_n]) \mid (a_1, \ldots, a_n) \in A_w \\ &\text{and there is some } (a'_1, \ldots, a'_n) \in R_A \\ &\text{with } ((a_1, \ldots, a_n), (a'_1, \ldots, a'_n)) \in \equiv_w \} \end{aligned}$$

and for $pop: s_1, \ldots, s_n \longrightarrow s \in POP$ analogously

$$\begin{aligned} \text{dom } pop_{A/\equiv} := \{ &([a_1], \ldots, [a_n]) \mid (a_1, \ldots, a_n) \in A_w \\ &\text{and there is some } (a'_1, \ldots, a'_n) \in \text{dom } pop_A \\ &\text{with } ((a_1, \ldots, a_n), (a'_1, \ldots, a'_n)) \in \equiv_w \} \end{aligned}$$

and if now $([a_1], \ldots, [a_n]) \in \text{dom } pop_{A/\equiv}$ with $(a'_1, \ldots, a'_n) \in \text{dom } pop_A$ and $((a_1, \ldots, a_n), (a'_1, \ldots, a'_n)) \in \equiv_w$, then

$$pop_{A/\equiv}([a_1], \ldots, [a_n]) := [pop_A(a'_1, \ldots, a'_n)]$$

Finally, for $op: s_1, \ldots, s_n \longrightarrow s \in OP$ and $(a_1, \ldots, a_n) \in A_w$

$$op_{A/\equiv}([a_1], \ldots, [a_n]) := [op_A(a_1, \ldots, a_n)]$$

In this way we get the *quotient* $A/\equiv$ of A by $\equiv$. Burmeister [Bur86, 2.6] proves that the construction is well-defined and $\text{nat}_\equiv: A \longrightarrow A/\equiv$ is a full surjective Σ-homomorphism with $\equiv = \ker \text{nat}_\equiv$.

Proposition 3.6.2 For all theories $T = (\Sigma, \Gamma)$ in $P(\stackrel{e}{=})$, coequalizers in $\mathbf{Mod}(T)$ are just quotients.

3.7. LEFT EXACT SKETCHES

Proof:
Surjective homomorphisms preserve existence equations by table 8.1 of [Bur86]. Therefore, the quotient taken in **Mod**(Σ) satisfies Γ. The coequalizing property now follows from the diagram completion lemma (Lemma 2.7.1 of [Bur86]). □

Proposition 3.6.3 For all theories $T = (\Sigma, \Gamma)$ in $P(\stackrel{e}{=})$, categorical intersections in **Mod**(T) are constructed by taking the relative subalgebra on the intersection of the carriers.

Proof:
By 4.2.4 of [Bur86], pullbacks in **Mod**(Σ) are constructed upon the pullbacks of the underlying sets. In particular, intersections are constructed by taking the relative subalgebra on the intersection of the carriers. Now if $t_1 \stackrel{e}{=} t_2 \in \Gamma$ and $\nu\colon X \longrightarrow A_1 \cap A_2$ is a valuation into the intersection,

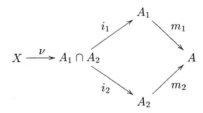

then we know that $(i_1 \circ \nu)^\#(t_1)$ and $(i_1 \circ \nu)^\#(t_2)$ are defined and equal, and similarly for i_2. Thus, assuming w. l. o. g. that i_1 and i_2 are inclusions, we have that $(i_1 \circ \nu)^\#(t_1) = (i_2 \circ \nu)^\#(t_1) = \nu^\#(t_1)$, and similarly for t_2. Thus $\nu^\#(t_1) = \nu^\#(t_2)$. □

Given a family of morphisms $(A \xrightarrow{h_i} B_i)_{i=1,\ldots,n}$, let $<h_1,\ldots,h_n>\colon A \longrightarrow B_1 \times \cdots \times B_n$ denote the unique morphism h with $A \xrightarrow{h} B_1 \times \cdots \times B_n \xrightarrow{\pi_i} B_i = A \xrightarrow{h_i} B_i$ for $i = 1,\ldots,n$.

3.7 Left exact sketches

The institution $LESKETCH$ (left exact sketches, see [Gra87, BW85]) has signatures $\Sigma = (G, U)$, where $G = (V, E, E \xrightarrow{start} V, E \xrightarrow{end} V)$ is a (finite) directed graph (with the possibility of multiple edges between two nodes) and U is a map assigning to each vertex (or object, to keep close to categorical terminology) a in V an edge (or arrow) $U(a) \in E$ from a to a. Signature morphisms $\sigma\colon (G, U) \longrightarrow (G', U')$ are graph homomorphisms $\sigma\colon G \longrightarrow G'$ such that

$\sigma \circ U = U' \circ \sigma$.

Let $|\mathbf{Set}|$ be the signature which is the underlying graph of the category of sets with U assigning to each set the identity function on that set. Then a Σ-model is a signature morphism $M\colon \Sigma \longrightarrow |\mathbf{Set}|$ and a model morphism is a natural transformation $\eta\colon M \longrightarrow M'$ (note that this is well defined because the target of M and M' is a category). If $\sigma\colon \Sigma \longrightarrow \Sigma'$ is a signature morphism, then $\mathbf{Mod}(\sigma)$ is the functor $(\_\circ \sigma)$ given by composing with σ.

A (G, U)-sentence is either

(1) a finite diagram in G, that is, a graph homomorphism $D\colon I \longrightarrow G$ (where I is some finite index graph), or

(2) a finite cone over a finite diagram $D\colon I \longrightarrow G$ in G, that is, an object L in G and arrows $\pi_i\colon L \longrightarrow D_i$ in G for $i \in |I|$.

Sentences are translated along σ by composing with σ, that is, $sen\,\sigma = (\sigma \circ \_)$.

A model $M\colon \Sigma \longrightarrow |\mathbf{Set}|$ satisfies a diagram, if M takes it to a commutative diagram in $\mathbf{Set}$, and M satisfies a cone, if M takes it to a limiting cone in $\mathbf{Set}$.

$LESKETCH$-theories are called left exact sketches (see [Gra87]). □

3.8 Order-sorted algebra with sort constraints

$COS(=:\Rightarrow=:)$ is the institution of coherent order sorted signatures, algebras and theories with conditional sort constraints as introduced by Goguen and Meseguer [GM92, GJM85].

Order-sorted signatures are triples $\Sigma = (S, \leq, OP)$, where $\leq$ is a partial order on S and (S, OP) is a many-sorted signature, such that the following *monotonicity condition* is satisfied:

$$op \in OP_{w,s} \cap OP_{w',s'} \text{ and } w \leq w' \text{ imply } s \leq s'.$$

Note that this condition excludes overloaded constants, so order-sorted signatures are *not* a generalization of many-sorted ($=\Rightarrow=$)-signatures.

A signature is called *regular*, if for each $op \in OP_{w,s}$ and $w_0 \leq w \in S^n$, there is a least arity for op that is greater than or equal to w_0. This condition implies that there also is a least *rank* (i. e. arity and coarity) for op with arity greater than or equal to w_0, see [GM92].

3.8. ORDER-SORTED ALGEBRA WITH SORT CONSTRAINTS

A signature is called *coherent*, if it is regular and locally filtered. The latter means that each connected component of $(S, \leq)$ must be filtered, i. e. any two elements (of the connected component) have a common upper bound.

Signature morphisms $\sigma\colon (S, \leq, OP) \longrightarrow (S', \leq', OP')$ are many-sorted signature morphisms $\sigma\colon (S, OP) \longrightarrow (S', OP')$ which satisfy the additional properties

1. $s_1 \leq s_2$ implies $\sigma^S(s_1) \leq' \sigma^S(s_2)$ (monotonicity)

2. $op \in OP_{w,s} \cap OP_{w',s'}$ implies $\sigma^{OP}_{w,s}(op) = \sigma^{OP}_{w',s'}(op)$ (preservation of overloading)

A signature morphism $\sigma\colon (S, \leq, OP) \longrightarrow (S', \leq', OP')$ between regular signatures is said to *preserve regularity*, if for each $op \in OP_{w,s}$, $w = s_1 \ldots s_n$, and $w_0 \leq w$, σ^{S^n} applied to the least arity for op which is $\geq w_0$ yields the least arity for $\sigma^{OP}_{w,s}(op)$ which is $\geq \sigma^{S^n}(w_0)$.

$(S, \leq, OP)$-models are (S, OP)-models A in $(\Longrightarrow=)$, such that

1. $s \leq s'$ implies $A_s \subseteq A_{s'}$

2. $op \in OP_{w,s} \cap OP_{w',s'}$ and $w \leq w'$ imply $op_A\colon A_w \longrightarrow A_s$ equals $op_A\colon A_{w'} \longrightarrow A_{s'}$ on A_w (monotonicity).

Homomorphisms $h\colon A \longrightarrow B$ are many-sorted homomorphisms $h\colon A \longrightarrow B$ in $(\Longrightarrow=)$ such that $s \leq s'$ and $a \in A_s$ imply $h_s(a) = h_{s'}(a)$. Reducts are defined as in $(\Longrightarrow=)$. It has been shown by Haxthausen and Nickl [HN96] that since a signature morphism preserves overloading, the corresponding reduct functor preserves monotonicity of algebras, while this is not the case for signature morphisms without the condition of preservation of overloading.

Given an S-sorted variable system X, the set $T_\Sigma(X)_s$ of $\Sigma(X)$-*terms of sort s* is inductively defined:

1. $x \in T_\Sigma(X)_s$ for $x \in X_s$

2. "$op(t_1, \ldots, t_n)$" $\in T_\Sigma(X)_s$ for $op\colon s_1, \ldots, s_n \longrightarrow s \in OP$ and $t_i \in T_\Sigma(X)_{s_i}$, $i = 1, \ldots, n$.

3. $T_\Sigma(X)_s \subseteq T_\Sigma(X)_{s'}$ for $s \leq s'$.

Again, it is straightforward to impose an algebraic structure on $T_\Sigma(X)$.

Note that, unlike in the many-sorted case, we do not have sort annotations here. Because of overloading and of the last clause, a term can have many different sorts. But we have the following:

Proposition 3.8.1 For regular signatures Σ, let $\Sigma^{\#}$ be the many sorted ($=\!\!\Rightarrow\!\!=$)-signature consisting of sort set S, an operation symbol $op_{s_1,\ldots,s_n,s}\colon s_1,\ldots,s_n \longrightarrow s$ for $op\colon s_1,\ldots,s_n \longrightarrow s \in OP$ plus additional coercion operation symbols $c_{s,s'}\colon s \longrightarrow s'$ whenever $s \leq s'$. Then each term $t \in T_{\Sigma}(X)$ has a *least sort* $LS_{\Sigma}(t)$ [GM92, 2.10] and a *least sort parse* $LP_{\Sigma}(t) \in T_{\Sigma^{\#}}(X)$ [GM92, p. 252]. If we restrict ourselves to signature morphisms preserving regularity, both constructions are natural in Σ.

Proof:
These are defined inductively as follows: If $t = x$ is a variable, it has a unique sort $LS_{\Sigma}(x)$ due to the disjointness condition for variable systems, and $LP_{\Sigma}(x) = x$. If $t = op(t_1,\ldots,t_s) \in T_{\Sigma}(X)_s$, then by induction hypothesis, t_i has least sort s_i. Take $w_0 = s_1\ldots s_n$. Then by the term formation rules, $op \in OP_{w',s'}$ for some w', s' with $s' \leq s$ and $w_0 \leq w'$. By regularity, there is a least rank w', s' for op such that $w' \geq w_0$. This least s' is the desired least sort of t, while $c_{s',s}(op_{w',s'}(c_{s_1,w_1'}(LP_{\Sigma}(t_1)),\ldots,c_{s_n,w_n'}(LP_{\Sigma}(t_n))))$ is the least sort parse of t.

Since signature morphisms preserve regularity, LS and LP are natural in Σ. □

The definition of sentences over signatures containing sort constraints in [GJM85] and [Yan93] is quite complex: The parse of terms has as result an annotated term plus some sort constraint conditions, which lead to only partial homomorphic extensions of assignments. While this resembles satisfaction in partial algebras, it is not clear to me at all if the satisfaction condition holds in this case, and in [Yan93], it is proved only for a restricted case: A simpler way to get an institution is to put the conditional sort constraints not into the signatures, but into the sentences. Of course, the effect of having more well-typed terms due to sort constraints is lost.

Let $COS(=\!:\!\Rightarrow\!=\!:)$ be the institution with coherent signatures, signature morphisms preserving regularity, order-sorted algebras as models, and Σ-sentences being conditional formulas

$$\forall X \,.\, e_1 \wedge \ldots \wedge e_n \Longrightarrow e$$

where X is an S-sorted system of variables and the atomic formulas e and e_i are of two kinds: Either

$$t_1 \stackrel{e}{=} t_2$$

where t_1, t_2 are Σ-terms in variables X with $LS_{\Sigma}(t_1)$ and $LS_{\Sigma}(t_2)$ in the same connected component of $(s, \leq)$. Or

$$t : s$$

where t is a Σ-term in the variables X and s is a sort in the same connected component as $LS_{\Sigma}(t)$. Sentence translation is defined inductively as usual.

A Σ-algebra A *satisfies* a conditional axiom, if all valuations that satisfy the premises also satisfy the conclusion. Satisfaction of atomic equations is defined as usual, and a valuation $\nu\colon X \longrightarrow A$ satisfies an atomic formula $t:s$ if

$$\nu^{\#}(t) \in A_s$$

where $\nu^{\#}$ is the homomorphic extension of ν to $T_\Sigma(X)$, see [GM92].

It follows from the results of [Yan93] that this is an institution.

Let us now define some restrictions.

A signature has a Noetherian ordering on the sorts if there is no infinite strictly increasing sequence of sorts. For these signatures, each connected component of sorts has a maximum element (Proposition 2.20 of [GM92]). This restriction is not severe: every finite coherent order-sorted signature is Noetherian as well. More severe is the further requirement that signature morphisms preserve these maximal elements of connected components. Together with monotonicity this implies that several connected components are joined into one if and only if their maximal elements are identified. This excludes as a signature morphism, for example, the inclusion of a specification of natural numbers into a specification of natural numbers as a subsort of integers.

Let $NCOS(=:\Rightarrow=:)$ be the thus outlined restriction of $COS(=:\Rightarrow=:)$. The advantage of $NCOS(=:\Rightarrow=:)$ over $COS(=:\Rightarrow=:)$ is that there is a much simpler embedding into $RP(R \stackrel{e}{=\!\Rightarrow} R \stackrel{e}{=})$, see Chapter 5, and that $NCOS(=:\Rightarrow=:)$ has composable signatures, see Proposition 3.9.3.

Further, we can restrict $COS(=:\Rightarrow=:)$ to conditional equations (that is, no sort constraints) and get $COS(=\Rightarrow=)$. The restriction to unconditional equations is denoted by $COS(=)$, while the restriction to unconditional equations and sort constraints is denoted by $COS(=:)$.

3.9 Composable signatures and amalgamation

Proposition 3.9.1 $RP(R =\!\Rightarrow R =)$ and all its restrictions have composable signatures.

Proof:
This is an easy generalization of the proof in [TBG91] that $(=)$ has a cocomplete signature category. □

Proposition 3.9.2 *LESKETCH* has composable signatures.

Proof:

The signature category of $LESKETCH$ is just the model category of the following (=)-theory:

sorts V, E
ops $start, end: E \longrightarrow V$
$\quad\quad U: V \longrightarrow E$
$\forall v : V .\ start(U(v)) = v$
$\forall v : V .\ end(U(v)) = v$

This model category is locally finitely presentable by Corollary 5.1.6 and thus cocomplete by Definition A.3.2. □

For $COS(=:\Rightarrow=:)$, I do not know whether the signature category has multiple pushouts or not. Haxthausen and Nickl [HN96] show that the inclusion of the category of coherent order-sorted signatures into the category of order-sorted signatures without the monotonicity and coherence condition does not preserve pushouts. In general, the theory of order-sorted signature colimits is complicated by the interplay between the ordering on the sorts and overloading of operation symbols (see [Mos98] for a more detailed study of this topic).

However, for the signature category of $NCOS(=:\Rightarrow=:)$, where for each property of the signatures there is a corresponding preservation property for signature morphisms, it is possible to algebraize the signature category and thus use a general cocompleteness theorem.

Proposition 3.9.3 $NCOS(=:\Rightarrow=:)$ has composable signatures.

Proof:
Consider the following specifications of signature categories as $P(\stackrel{e}{=}\Rightarrow\stackrel{e}{=})$-theories (where $D(t)$ abbreviates $t \stackrel{e}{=} t$):

spec SIG =
  sorts S, S^*, OP
  ops $\lambda: \longrightarrow S^*$
  $\quad\quad \_\_\_\_ : S \times S^* \longrightarrow S^*$
  preds $\_\_ : \_\_ \longrightarrow \_\_ : OP \times S^* \times S$

spec OSASIG = SIG then
  preds $\leq : S \times S$
  $\forall s : S .\ s \leq s$ %(reflexivity)%
  $\forall s_1, s_2, s_3 : S .\ s_1 \leq s_2 \wedge s_2 \leq s_3 \Longrightarrow s_1 \leq s_3$ %(transitivity)%
  $\forall s_1, s_2 : S .\ s_1 \leq s_2 \wedge s_2 \leq s_1 \Longrightarrow s_1 = s_2$ %(antisymmetry)%
  $\forall op : OP, s_1, \ldots, s_n, s, s'_1, \ldots, s'_n, s' : S .$
  $\quad op: s_1, \ldots, s_n \longrightarrow s \wedge op: s'_1, \ldots, s'_n \longrightarrow s' \wedge \bigwedge_{i=1, \ldots, n} s_i \leq s'_i \Longrightarrow s \leq s'$
  $\quad\quad$ for $n \in I\!\!N$ %(monotonicity)%

3.9. COMPOSABLE SIGNATURES AND AMALGAMATION

In the sequel, let lr_k^n abbreviate $leastrank_k^n(op, s_1, \ldots, s_n, s_0, s'_1, \ldots, s'_n)$.

spec REGOSASIG = OSASIG then
  ops $leastrank_k^n : OP \times S^{2n+1} \longrightarrow ?S$ for $n \in \mathbb{N}$, $k = 0, \ldots, n$
  $\forall op\!:\! OP, s_1, \ldots, s_n, s_0, s'_1, \ldots, s'_n\!:\! S$.
  $\qquad D(lr_k^n) \iff (op\!:\! s_1, \ldots, s_n \longrightarrow s_0 \wedge \bigwedge_{i=1,\ldots,n} s'_i \leq s_i)$
  $\qquad\qquad\qquad$ for $n \in \mathbb{N}$, $k = 0, \ldots, n$
  $\forall op\!:\! OP, s_1, \ldots, s_n, s_0, s'_1, \ldots, s'_n\!:\! S$. $D(lr_k^n) \implies s'_k \leq lr_k^n$
  $\qquad\qquad\qquad$ for $n \in \mathbb{N}$, $k = 0, \ldots, n$
  $\forall op\!:\! OP, s_1, \ldots, s_n, s_0, s'_1, \ldots, s'_n\!:\! S$. $D(lr_0^n) \implies op\!:\! lr_1^n, \ldots, lr_n^n \longrightarrow lr_0^n$
  $\qquad\qquad\qquad$ for $n \in \mathbb{N}$
  $\forall op\!:\! OP, s_1, \ldots, s_n, s_0, s'_1, \ldots, s'_n, s''_1, \ldots, s''_n, s''_0\!:\! S$.
  $\qquad D(lr_0^n) \wedge op\!:\! s''_1, \ldots, s''_n \longrightarrow s''_0 \wedge \bigwedge_{i=1,\ldots,n} s'_i \leq s''_i \implies lr_k^n \leq s''_k$
  $\qquad\qquad\qquad$ for $n \in \mathbb{N}$, $k = 0, \ldots, n$

spec NCOSASIG = REGOSASIG then
  ops $max\!:\! S \longrightarrow S$
  $\forall s\!:\! S$. $s \leq max(s)$
  $\forall s_1, s_2\!:\! S$. $max(s_1) \leq s_2 \implies max(s_1) = s_2$
  $\forall s_1, s_2\!:\! S$. $s_1 \leq s_2 \implies max(s_1) = max(s_2)$

Now **Mod(OSASIG)** rather directly specifies the category of order-sorted signatures and signature morphisms preserving overloading (but not necessarily regularity), with the one difference that we use a global space of operation symbols. By Fact 3 of [HN96], such a global treatment of operation symbols is equivalent to the standard local one, if the global space OP of operation symbols is the union of all local sets $OP_{w,s} := \{\, op \in OP \mid op\!:\! w \longrightarrow s \,\}$. Now **OSASIG** does not exclude elements $op \in OP$ for which there is no w, s with $op\!:\! w \longrightarrow s$. But if we take the full subcategory of **Mod(OSASIG)** consisting of those algebras where

- each $op \in OP$ is "non-junk", that is, is related to some w, s via $op\!:\! w \longrightarrow s$, and

- S^* is generated by S, $empty$ and add,

then this is isomorphic to the category of order-sorted signatures and signature morphisms preserving overloading.

Now given an operation symbol $op\!:\! w \longrightarrow s$ and $w_0 \leq w$, the operation $leastrank$ of **REGOSASIG** can be seen to yield the least rank for op with arity $\geq w_0$ as $leastrank_1^n(op, w, s, w_0), \ldots, leastrank_n^n(op, w, s, w_0), leastrank_0^n(op, w, s, w_0)$. The first axiom schema just states the conditions $op\!:\! w \longrightarrow s$ and $w_0 \leq w$ as definedness conditions for $leastrank$, the second axiom schema states that *leastrank* delivers a rank $\geq w_0$. The third axiom schema states that *leastrank* delivers a rank of *op*, and the forth axiom schema states that this rank is the least rank of *op* with arity $\geq w_0$. Thus the full subcategory of **Mod(REGOSASIG)** consisting of signatures with only non-junk operation symbols and non-junk sort sequences is isomorphic to the category of regular signatures and signature morphisms preserving regularity.

Likewise, the full subcategory of **Mod(NCOSASIG)** consisting of signatures with only non-junk operation symbols and non-junk sort sequences can be seen to be isomorphic to the category of Noetherian coherent signatures and signature morphisms preserving regularity and maximal elements of connected components.

Now **Mod(NCOSASIG)** is locally finitely presentable by Corollary 5.1.6 and thus cocomplete by Definition A.3.2. The full subcategory consisting of signatures with only non-junk operation symbols and non-junk sort sequences is coreflective: the coreflector just deletes all junk operation symbols and all junk sort sequences. The coreflection condition is shown by noting that a signature morphism $\sigma: \Sigma \longrightarrow \Sigma'$ from a signature without junk operation symbols and without junk sort sequences can be viewed as signature morphism into the coreflection of Σ' since $op: w \longrightarrow s$ implies $\sigma(op): \sigma(w) \longrightarrow \sigma(s)$, so all operation symbols in the image are non-junk, and a similar argument holds for sort sequences.

By Proposition A.2.2, the full subcategory of **Mod(NCOSASIG)** consisting of signatures with only non-junk operation symbols is cocomplete as well, and so is the signature category of $NCOS(=:\Rightarrow=:)$. □

Concerning amalgamation, Claßen, Große-Rhode and Wolter [CGRW95] prove the following folklore theorem [BW85, Tar86b]:

Theorem 3.9.4 The specification frame $\mathbf{Th_0}(P(\stackrel{e}{=}\Rightarrow\stackrel{e}{=}))$ has amalgamation.
□

By Proposition 2.3.4, this carries over to the institution $P(\stackrel{e}{=}\Rightarrow\stackrel{e}{=})$ and by proposition 4.7.8 to the subinstitutions of $P(\stackrel{e}{=}\Rightarrow\stackrel{e}{=})$. Later, we will see that most of the remaining institutions have amalgamation as well (Corollary 5.1.3).

3.10 Liberality of some institutions

In section 2.7, we introduced Parameterized Abstract Data Types and liberal institutions. We now show some of the above introduced institutions to be

3.10. LIBERALITY OF SOME INSTITUTIONS

liberal. Later on, we will show how this carries over to the other institutions (see corollary 5.1.4).

Fact 3.10.1 In $RP(R \stackrel{e}{=}\Rightarrow R \stackrel{e}{=})$ and all its restrictions, for each parameterized theory $T \stackrel{\theta}{\longrightarrow} T1$, a free PADT exists. We denote it by (η^θ, F^θ). Thus $RP(R \stackrel{e}{=}\Rightarrow R \stackrel{e}{=})$ and all its restrictions are liberal.

Proof: This follows from a general theorem of Tarlecki [Tar85]. Since we need a definite construction later on, we follow Reichel's book ([Rei87], 3.2, 3.3.1 and 3.5.6.) and [EM85]. Let $T \stackrel{\theta}{\longrightarrow} T1$ be a parameterized theory and A a T-model. $F^\theta(A)$ can be constructed as follows: Let $T1 = (\Sigma 1, \Gamma 1)$, $\Sigma 1(A)$ be $\Sigma 1$ augmented by total constants $c^a : \theta\, s$ for $a \in A_s$, and let $\Gamma 1(A)$ be $\Gamma 1$ plus axioms $\theta\, op(c^{a_1, \ldots, a_n}) \stackrel{e}{=} c^{op_A(a_1, \ldots, a_n)}$ for $(a_1, \ldots, a_n) \in \mathrm{dom}\, op_A$ and $op \in OP \cup POP$ and $\theta\, R(c^{a_1, \ldots, a_n})$ for $(a_1, \ldots, a_n) \in R_A$, $R \in REL$. We write $T_{\Sigma 1}(A)$ for $T_{\Sigma 1(A)}\,|_{\Sigma 1 \hookrightarrow \Sigma 1(A)}$. Putting $T1(A) = (\Sigma 1(A), \Gamma 1(A))$, let $H_{T1}(A) \subseteq T_{\Sigma 1}(A)$ be the algebra of all terms for which $T1(A) \models t \stackrel{e}{=} t$ with usual term algebra operations and relations $R_{H_{T1}(A)} = \{\, t_1, \ldots, t_n \mid T1(A) \models R(t_1, \ldots, t_n)\,\}$. Then there is a quotient (i.e. a full surjection) $H_{T1}(A) \stackrel{q_A}{\longrightarrow} F^\theta(A)$ with kernel $\ker q_A = \{\, (t_1, t_2) \mid T1(A) \models t_1 \stackrel{e}{=} t_2\,\}$. □

Chapter 4

Different types of arrow between logical systems

> "It should be observed first that the whole concept of a category is essentially an auxiliary one; our basic concepts are essentially those of a functor and of a natural transformation." *S. Eilenberg and S. Mac Lane*[1]

In chapter 2, we considered the distinction between proof theory and model theory as the main argument for having several types of logical system. But there was another argument: To work just with abstract theories and having no notion of axiom, as specification frames do, is a very semantic point of view not amenable to theorem proving. In many contexts, we need institutions with their distinction between the notions of signature and of sentence. But there are many choices for what should be included into the signature part, and what into the sentences. For example, in type theories [HHP93], typing information is put into the sentences, while in HEP-theories [Rei87], not only typing, but also definedness axioms are put into the signatures.

The problem now arises when considering representations. Plain representations (of institutions or entailment systems), which keep signatures and sentences separated, occur in many contexts, but typically only as trivial sub-logical systems. However, there are many representations, which should even be called embeddings, that are not plain.

Meseguer's notion of simple institution representation (or entailment system) [Mes89] is a first solution: signatures are mapped to theories, while sentences

---

[1]Transactions of the American Mathematical Society 58 (1945).

are mapped to sentences. However, there are contexts, in which this does not suffice. Thus there is a need for notions of representations located between simple institution representations and specification frame representations.

Using monads and adjunctions, the well-known simple representations and three new notions of representations between institutions are introduced below. These vary in the strictness of keeping the signature–sentence distinction. In each case, we briefly show the application to different logical systems. More applications are presented in chapter 5.

A short version of this chapter has been published in [Mos96a]. The notion of embedding of institutions was published first in [KM95].

4.1 The theory monad and simple representations of institutions

We introduce Meseguer's simple institution representations of here not from scratch, but rather in a structured way using a monadic construction to enrich the target institution. Compared to plain representations, simple representations have a greater flexibility. When constructing such a representation, we may use the added expressiveness of the enriched target institution.

In Definition 2.2.4, we have introduced the functor $\mathbf{Th_0}\colon \mathbf{PlainInst} \longrightarrow \mathbf{SpecFram}$, which, for an institution $\mathcal{I}$, yields category of theories $\mathbf{Th_0}(\mathcal{I})$ and a model functor starting from this category. Now it is easy to define a sentence functor starting from $\mathbf{Th_0}(\mathcal{I})$ as well by just composing $sen\colon \mathbf{Sign} \longrightarrow \mathbf{Set}$ with $sign\colon \mathbf{Th_0}(\mathcal{I}) \longrightarrow \mathbf{Sign}$. Similarly, satisfaction w. r. t. to a theory T is just satisfaction w. r. t. $sign(T)$. Thus, we get a functor from $\mathbf{PlainInst}$ to $\mathbf{PlainInst}$, which by abuse of language, we also denote by $\mathbf{Th_0}$. Its action on representations is defined as follows: $\mathbf{Th_0}(\Phi, \alpha, \beta) = (\Phi^\alpha, \alpha_{sign}, \beta^{\mathbf{Th_0}})$.

There is an obvious inclusion representation $\eta_\mathcal{I}\colon \mathcal{I} \longrightarrow \mathbf{Th_0}(\mathcal{I})$, which maps Σ to $(\Sigma, \emptyset)$, and a projection $\xi_\mathcal{I}\colon \mathbf{Th_0}(\mathbf{Th_0}(\mathcal{I})) \longrightarrow \mathbf{Th_0}(\mathcal{I})$ mapping $((\Sigma, \Gamma), \Gamma')$ to $(\Sigma, \Gamma \cup \Gamma')$. Both are the identity for sentences and models.

Proposition 4.1.1 $(\mathbf{Th_0}, \eta, \xi)$ is a monad over $\mathbf{PlainInst}$.

Proof:
The identity laws follow from $\emptyset$ being an identity w. r. t. union, and the associativity law from union being associative. □

Definition 4.1.2 The monad $\mathbf{Th_0} = (\mathbf{Th_0}, \eta, \xi)$ over $\mathbf{PlainInst}$ induces the Kleisli category $\mathbf{Inst} = \mathbf{PlainInst}_{\mathbf{Th_0}}$ of institutions and simple institution

4.1. THE THEORY MONAD AND SIMPLE REPRESENTATIONS OF INSTITUTIONS

representations. □

Recall that objects of the Kleisli category are the same as those of **PlainInst**, but morphisms $\mu: \mathcal{I} \longrightarrow \mathcal{I}'$ are representations $\mu: \mathcal{I} \longrightarrow \mathbf{Th_0}(\mathcal{I}')$ in **PlainInst**, thus we are allowed to map a signature not just to a signature, but to a theory, which gives us more flexibility when setting up representations of institutions.

Also, we have the Kleisli adjunction

$$\mathbf{PlainInst} \underset{U_{\mathbf{Th_0}}}{\overset{F_{\mathbf{Th_0}}}{\rightleftarrows}} \mathbf{Inst}$$

where $F_{\mathbf{Th_0}}$ is just the inclusion (every plain representation is also a simple representation), while $U_{\mathbf{Th_0}}$ essentially is $\mathbf{Th_0}$.

A typical example is the simple institution representation from partial first-order logic to total first-order logic where partial functions are mapped to total functions plus a partial congruence relation, sentences are translated by replacing existential equality by the partial congruence and models are translated by factoring modulo the congruence. Here, $\Phi(\Sigma)$ has to contain the congruence axioms, for example. This translation allows to re-use the proof theory via borrowing, see [CMR99]. There are plenty of other simple representations, see [Cer93, CM93, AC92, Mes89] and chapter 5.

Using Corollary 2.3.6, we can repeat the whole section with **PlainInst** replaced by $\mathbf{PlainInst}^{amal}$. The only thing we need to prove is that for $I \in \mathbf{PlainInst}^{amal}$, η_I and ξ_I preserve composable signatures and amalgamation. For η_I, this is immediate. For ξ_I, which is of form (Φ, id, id), let $\mathcal{S} = (((\Sigma, \Gamma), \Gamma'), (((\Sigma, \Gamma), \Gamma') \xrightarrow{\sigma_i} ((\Sigma_i, \Gamma_i), \Gamma'_i))_{i \in I})$ be a source in the signature category of $\mathbf{Th_0}(\mathbf{Th_0}(\mathcal{I}))$. Then, in the notation of the proof of proposition 2.3.3, we can show that ξ_I preserves composable signatures:

$$\begin{aligned}
& \Phi(tip(Colim(\mathcal{S}))) \\
=\ & \Phi^{id}(tip(Colim(sign(\mathcal{S}))), id[\bigcup_{i \in I} \rho_i[\Gamma'_i]]) \\
=\ & (\Phi^{id})^{id}((tip(Colim(sign(sign(\mathcal{S})))), id[\bigcup_{i \in I} \rho_i[\Gamma_i]]), id[\bigcup_{i \in I} \rho_i[\Gamma'_i]]) \\
=\ & (tip(Colim(sign(sign(\mathcal{S})))), id[\bigcup_{i \in I} \rho_i[\Gamma_i] \cup id[\bigcup_{i \in I} \rho_i[\Gamma'_i]]) \\
=\ & (tip(Colim(sign(sign(\mathcal{S})))), \bigcup_{i \in I} \rho_i[\Gamma_i \cup \Gamma'_i]) \\
=\ & tip(Colim(\Phi(\mathcal{S})))
\end{aligned}$$

Preservation of amalgamation follows from proposition 2.3.7.

This gives us a category $\mathbf{Inst}^{amal}$ of institutions with amalgamation and simple institution representations preserving amalgamation.

Likewise, by enlarging the source and target of $\mathbf{Th_0}$ to **PlainInstrps**, we get a category **Instrps** of rps pre-institutions and simple rps pre-institution representations.

4.2 The conjunctive monad and conjunctive representations of institutions

Now simple representations of institutions cover more, but still not all desirable representations of institutions. A second step is to allow sentences being mapped not to single sentences, but to (finite) sets of sentences.

An institution $\mathcal{I} = (\mathbf{Sign}, sen, \mathbf{Mod}, \models)$ can be enriched to the institution $\bigwedge(\mathcal{I}) = (\mathbf{Sign}, sen^\wedge, \mathbf{Mod}, \models^\wedge)$, where $sen^\wedge = \mathcal{P}_{fin} \circ sen$, the composition of sen with the functor $\mathcal{P}_{fin}: \mathbf{Set} \longrightarrow \mathbf{Set}$ giving the set of finite subsets, and $M \models_\Sigma^\wedge S$ iff for all $\varphi \in S$, $M \models_\Sigma \varphi$.

Again, this construction can easily be turned into a monad $\mathbf{Conj} = (\bigwedge, \eta, \xi)$ (this time acting on **Inst**).

Definition 4.2.1 The morphisms of the Kleisli category $\bigwedge \mathbf{Inst} = \mathbf{Inst}_{\mathbf{Conj}}$ are called conjunctive simple representations of institutions. □

Conjunctive representations are not needed, of course, if the target institution already has conjunction available. But consider the representation of $P(\stackrel{s}{=})$ in $P(\stackrel{e}{\Rightarrow}\stackrel{e}{=})$: Here, a strong equation $t_1 \stackrel{s}{=} t_2$ is mapped to $(t_1 \stackrel{e}{=} t_1 \Longrightarrow t_1 \stackrel{e}{=} t_2) \wedge (t_2 \stackrel{e}{=} t_2 \Longrightarrow t_1 \stackrel{e}{=} t_2)$.

More complex examples are studied in chapter 5.

We can extend the re-use of theorem provers (cf. Theorem 2.6.1) to simple and to conjunctive representations:

Corollary 4.2.2 Let

1. $\mu = (\Phi, \alpha, \beta): \mathcal{I} \longrightarrow \mathcal{I}'$ be a conjunctive simple institution representation with surjective model components, or

2. let $\mathcal{I}$ be an institution in which model isomorphisms preserve satisfaction and let $\mu = (\Phi, \alpha, \beta): \mathcal{I} \longrightarrow \mathcal{I}'$ be a conjunctive institution representation with model components surjective up to isomorphism, i. e. for each $\Sigma \in \mathbf{Sign}$ and $M \in \mathbf{Mod}(\Sigma)$, there is an $M' \in \mathbf{Mod}'(\Phi(\Sigma))$ with $\beta_\Sigma(M') \cong M$.

Then
$$\Gamma \models_\Sigma^\mathcal{I} \varphi \text{ iff } \bigcup \alpha_\Sigma[\Gamma] \cup ax(\Phi(\Sigma)) \models_{sign(\Phi(\Sigma))}^{\mathcal{I}'} ax(\Phi(\Sigma)) \cup \alpha_\Sigma(\varphi)$$
where now $\alpha_\Sigma(\varphi)$ is a finite set of sentences.

□

Again, this section can be applied also to **Inst**^{amal} and **Instrps**.

4.3 The presentation extension monad and weak representations of institutions

Still, there is the need for even more complex representations of institutions. In many examples, it is only possible to map theories to theories. To get useful results, we have to restrict ourselves to theories with finite sets of sentences, i. e. to *presentations*.

Definition 4.3.1 A weak rps pre-institution representation $\mu: \mathcal{I} \longrightarrow \mathcal{I}'$ is a specification frame representation $\mu: \mathbf{Pres}(\mathcal{I}) \longrightarrow \mathbf{Pres}(\mathcal{I}')$. Composition is that of **SpecFram**. This gives us a category **WeakInstrps** of rps pre-institutions and weak rps pre-institution representations. The full subcategory generated by the institutions is denoted by **WeakInst** (institutions and weak institution representations).

Typical examples for weak representations of institutions are the following constructions:

- explicit definition of λ-abstraction: There is a weak representation from higher-order logic to higher-order logic without λ-abstraction. A HOL-formula is translated by recursively replacing each subterm $\lambda x_\alpha.t_\beta$ in a context $x_{1\alpha_1}, \ldots, x_{n\alpha_n}$ with $f x_{1\alpha_1} \ldots x_{n\alpha_n}$, where $f : \alpha_1 \to \ldots \to \alpha_n \to \alpha \to \beta$ is a new function symbol with an equation $f x_{1\alpha_1} \ldots x_{n\alpha_n} x_\alpha = t_\beta$ in the presentation extension.

- explicit definition of description operators: There is a weak representation from partial function higher-order logic with definite description operators [Far91] to partial function higher-order logic *without* description operators. A *PHOL*-formula is translated to a presentation extension containing the formula itself plus the axioms Δ_α from [Far91] for each description operator $d_{(\alpha \to *) \to \alpha}$ contained in the formula.

Further examples are studied in Chapter 5.

In some cases, a weak representation can be shown to be modular. That means, the mapping of presentations to presentations can be split into two maps: one mapping signatures to presentations, and the other mapping sentences to extensions of these presentations. Now this construction is not just a monad as in the previous sections, but comes very naturally via an adjunction between specification frames and institutions. Unfortunately, there seems to be no adjunction between specification frames and institutions in general. We either have to enlarge or restrict both categories.

We begin with the case of enlargement. (Strictly speaking, we first slightly restrict things, namely to composable signatures resp. presentations.) We are looking for an adjoint to **Pres: Instrps** $\longrightarrow$ **SpecFram**, which adds sentences to specification frames in a canonical way. One idea is to consider, for a specification frame $\mathcal{F}$, presentation extensions $\theta: T \longrightarrow T_1$ as sentences of an institution. A presentation extension should not be an arbitrary presentation morphism, but its reduct functor should actually restrict the models of the presentation, since this, after all, is the crucial property of sentences. Technically, $\mathbf{Mod}(\theta)$ shall be injective on objects. But to make the adjoint functorial using these presentation extensions, we need representations to preserve amalgamation. Therefore, we defer the requirement of this injectivity property to the case with amalgamation.

Proposition 4.3.2 The functor **Pres: Instrps**^{comp} $\longrightarrow$ **SpecFram**^{comp} has a right adjoint, the presentation extension functor **derive**. With the data given below, we have an adjoint situation (η, ϵ): **Pres** $\dashv$ **derive**: **SpecFram**^{comp} $\longrightarrow$ **Instrps**^{comp}.

Proof:
For a specification frame $\mathcal{F} = (\mathbf{Th}, \mathbf{Mod})$, we set $\mathbf{derive}(\mathcal{F}) = (\mathbf{Th}, sen, \mathbf{Mod}, \models)$, where $sen(T)$ is the set of all sentences

$$\textbf{derive from } T_1 \textbf{ by } \theta$$

with $\theta: T \longrightarrow T_1$ a theory morphism in **Th**.

A sentence **derive from** T_1 **by** θ is translated along a signature morphism in $\mathbf{derive}(\mathcal{F})$, i. e. a theory morphism in $\mathcal{F}$, $\sigma: T \longrightarrow T'$, to **derive from** T_1' **by** θ' given by taking the pushout

$$\begin{array}{ccc} T & \xrightarrow{\sigma} & T' \\ \theta \downarrow & & \downarrow \theta' \\ T_1 & \xrightarrow{\sigma_1} & T_1' \end{array}$$

Now $M \models \textbf{derive from } T' \textbf{ by } \sigma$ iff there is some $M' \in \mathbf{Mod}(T')$ with $M'|_\sigma = M$. We have to show the rps-half of the satisfaction condition. Let $\sigma: T \longrightarrow T'$ be a $\mathbf{derive}(\mathcal{F})$-signature morphism, and let $M' \models_{T'} \textbf{derive from } T_1' \textbf{ by } \theta'$.

4.3. THE PRESENTATION EXTENSION MONAD AND WEAK REPRESENTATIONS OF INSTITUTIONS

This means that there is some $M_1' \in \mathbf{Mod}(T_1')$ with $M_1'|_{\theta'} = M'$. By putting $M_1 := M_1'|_{\sigma_1} \in \mathbf{Mod}(T_1)$, we have $M_1|_{\theta} = M_1'|_{\sigma_1 \circ \theta} = M_1'|_{\theta' \circ \sigma} = M'|_{\sigma}$, i. e. M_1 is the witness for $M'|_{\sigma} \models_T$ **derive from** T_1 **by** θ.

derive($\mathcal{F}$) having composable signatures follows immediately from $\mathcal{F}$ having composable theories.

For morphisms, **derive**(Φ, β) = (Φ, Φ, β). We have to show the rps-half of the representation condition. Let $M' \models_{\Phi(T)}$ **derive from** $\Phi(T_1)$ **by** $\Phi(T) \xrightarrow{\Phi(\theta)} \Phi(T_1)$. We can now repeat the above argument with $\mathbf{Mod}(\sigma)$ replaced by β_T and $\mathbf{Mod}(\sigma_1)$ replaced by β_{T_1} to show that $\beta_T(M') \models_T$ **derive from** T_1 **by** $T \xrightarrow{\theta} T_1$.

Again, **derive**(Φ, β) preserving composable signatures follows from (Φ, β) preserving composable theories.

The unit $\eta_\mathcal{I} : \mathcal{I} \longrightarrow$ **derive**(**Pres**($\mathcal{I}$)) of the adjunction is given by mapping a signature Σ to the presentation $(\Sigma, \emptyset)$, and mapping a Σ-sentence φ to the $(\Sigma, \emptyset)$-sentence

$$\textbf{derive from } (\Sigma, \{\varphi\}) \textbf{ by } \sigma$$

where σ is the inclusion of $(\Sigma, \emptyset)$ into $(\Sigma, \{\varphi\})$, and leaving models unchanged. The representation condition, preservation of composable signatures and naturality of η are shown easily.

The counit $\epsilon_\mathcal{F} : \mathbf{Pres}(\mathbf{derive}(\mathcal{F})) \longrightarrow \mathcal{F}$ of the adjunction is defined as (Φ, β):

- For a presentation $TT = (T, \Gamma)$, take $\Phi(T) := T_0$, where $(T_0, \rho : T \longrightarrow T_0, (\rho_\theta : T_\theta \longrightarrow T_0)_{\textbf{derive from } T_\theta \textbf{ by } T \xrightarrow{\theta} T_\theta \in \Gamma}) = Colim(\mathcal{T})$ with $\mathcal{T} = (T, (T \xrightarrow{\theta} T_\theta)_{\textbf{derive from } T_\theta \textbf{ by } T \xrightarrow{\theta} T_\theta \in \Gamma})$.

- Let $\sigma : TT \longrightarrow TT'$ (where $TT' = (T', \Gamma')$) be a presentation morphism, that is, $\sigma : T \longrightarrow T'$ and for each **derive from** T_θ **by** $T \xrightarrow{\theta} T_\theta \in \Gamma$, we have a **derive from** $T'_{\theta'}$ **by** $T' \xrightarrow{\theta'} T'_{\theta'} \in \Gamma'$ with the square in

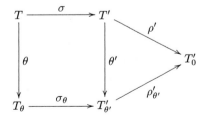

being a pushout. Let

$$(T_0, \rho : T \longrightarrow T_0, (\rho_\theta : T_\theta \longrightarrow T_0)_{\textbf{derive from } T_\theta \textbf{ by } T \xrightarrow{\theta} T_\theta \in \Gamma})$$

be the multiple pushout of $\mathcal{T}$ as above, and similarly for $\mathcal{T}'$ constructed out of TT'. Since the above triangle commutes as well as the square, $(T, T \xrightarrow{\sigma} T' \xrightarrow{\rho'} T'_0, (T_\theta \xrightarrow{\sigma_\theta} T'_{\theta'} \xrightarrow{\rho'_{\theta'}} T'_0)_{\text{derive from } T_\theta \text{ by } T \xrightarrow{\theta} T_\theta \in \Gamma})$ is a cocone over $\mathcal{T}$. This cocone induces a presentation morphism from T_0 to T'_0, which we take as $\Phi(\sigma)$.

- Preservation of composable theories can be shown by a rather complex chasing along diagrams, while using commutation of colimits (dual of [HS73, 25.4]).

- β_{TT} is just $\mathbf{Mod}(\rho): \mathbf{Mod}(T_0) \longrightarrow \mathbf{Mod}(T)$.

- Naturality of ϵ follows from Φ preserving multiple pushouts and naturality of β.

To show adjointness, we must prove that

$$\mathbf{derive}(\mathcal{F}) \xrightarrow{\eta_{\mathbf{derive}(\mathcal{F})}} \mathbf{derive}(\mathbf{Pres}(\mathbf{derive}(\mathcal{F}))) \xrightarrow{\mathbf{derive}(\epsilon_\mathcal{F})} \mathbf{derive}(\mathcal{F})$$
$$= id_{\mathbf{derive}(\mathcal{F})}$$

and

$$\mathbf{Pres}(\mathcal{I}) \xrightarrow{\mathbf{Pres}(\eta_\mathcal{I})} \mathbf{Pres}(\mathbf{derive}(\mathbf{Pres}(\mathcal{I}))) \xrightarrow{\epsilon_{\mathbf{Pres}(\mathcal{I})}} \mathbf{Pres}(\mathcal{I}) = id_{\mathbf{Pres}(\mathcal{I})}$$

The former equation is almost trivial for signatures and models. For sentences, we have that **derive from** T_1 **by** $T \xrightarrow{\theta} T_1$ is mapped by $\eta_{\mathbf{derive}(F)}$ to **derive from** $(T, \{ \mathbf{derive\ from\ } T_1 \mathbf{\ by\ } T \xrightarrow{\theta} T_1 \})$ **by** ι, where ι is the inclusion of $(T, \emptyset)$ into $(T, \{ \mathbf{derive\ from\ } T_1 \mathbf{\ by\ } T \xrightarrow{\theta} T_1 \})$. This in turn is mapped by $\mathbf{derive}(\epsilon_F)$ to the multiple pushout of $(T, T \xrightarrow{\theta} T_1)$, which again is **derive from** T_1 **by** $T \xrightarrow{\theta} T_1$.

The latter equation for presentations: (Σ, Γ) in $\mathbf{Pres}(\mathcal{I})$ is mapped by $\mathbf{Pres}(\eta_I)$ to $((\Sigma, \emptyset), \{ \mathbf{derive\ from\ } (\Sigma, \{ \varphi \}) \mathbf{\ by\ } (\Sigma, \emptyset) \hookrightarrow (\Sigma, \{ \varphi \}) \mid \varphi \in \Gamma \})$. $\epsilon_{\mathbf{Pres}(\mathcal{I})}$ maps this to the (tip of the) colimit of this source, which is (Σ, Γ). Concerning models, a (Σ, Γ)-model M is mapped by $\epsilon_{\mathbf{Pres}(\mathcal{I})}$ to $M \mid_{(\Sigma, \emptyset) \longrightarrow (\Sigma, \Gamma)}$ which is M. $\mathbf{Pres}(\eta_\mathcal{I})$ leaves this unchanged as well. $\square$

Now the adjunction induces a monad **Ext-rps** $=$ (**derive** $\circ$ **Pres**, η, **derive**($\epsilon_{\mathbf{Pres}}$)) on $\mathbf{Instrps}^{comp}$ with Kleisli category $\mathbf{Instrps}^{comp}_{\mathbf{Ext\text{-}rps}}$.

Proposition 4.3.3
The functor $K^{rps}: \mathbf{Instrps}^{comp}_{\mathbf{Ext\text{-}rps}} \longrightarrow \mathbf{WeakInstrps}^{comp}$ being the identity on objects and $K_{\mathbf{Ext\text{-}rps}}$ on morphisms is an isomorphism.

Proof:
$K_{\textbf{Ext-rps}}$ is full and faithful (see A.1). □

Before interpreting this result, we first shift it to the level of institutions. To be able to do this, we have to restrict ourselves to institutions with amalgamation.

Proposition 4.3.4 The functor $\textbf{Pres} \colon \textbf{Inst}^{amal} \longrightarrow \textbf{SpecFram}^{amal}$ has a right adjoint, denoted by **derive!**. With the data given below, we have an adjoint situation $(\eta, \epsilon) \colon \textbf{Pres} \dashv \textbf{derive!} \colon \textbf{SpecFram}^{amal} \longrightarrow \textbf{Inst}^{amal}$.

Proof:
We have to redefine **derive** from Proposition 4.3.2, while η and ϵ can be kept. Let $\textbf{derive!}(\mathcal{F})$ inherit signatures and models from $\textbf{derive}(\mathcal{F})$, and sentences are restricted to those sentences

$$\text{derive from } T' \text{ by } \sigma$$

such that $M'_1 \mid_\sigma = M'_2 \mid_\sigma \Rightarrow M'_1 = M'_2$. Again, sentence translation is defined by pushing out. As before, $M \models_T \text{derive from } T_1 \text{ by } T \xrightarrow{\theta} T_1$ if there is a T_1-expansion (which now automatically is unique by the above condition) of M.

Due to the semantic condition that (reduct functors corresponding to) sentences have to satisfy, the set of sentences here is not syntactically determined. But at least, some subset, namely extensions by definitions, can be determined syntactically, since extensions by definitions admit exactly one model expansion.

$\textbf{derive!}(\mathcal{F})$ satisfies eps: Let $\text{derive from } T_1 \text{ by } \theta$ be a T-sentence in $\textbf{derive!}(\mathcal{F})$, which is translated along a signature morphism $\sigma \colon T \longrightarrow T'$ to $\text{derive from } T'_1 \text{ by } \theta'$, given by the pushout

$$\begin{array}{ccc} T & \xrightarrow{\sigma} & T' \\ {\scriptstyle \theta} \downarrow & & \downarrow {\scriptstyle \theta'} \\ T_1 & \xrightarrow{\sigma_1} & T'_1 \end{array}$$

Let $M' \in \textbf{Mod}(T')$ with $M' \mid_\sigma \models_T \text{derive from } T_1 \text{ by } T \xrightarrow{\theta} T_1$. This means that there is a T_1-model M_1 with $M_1 \mid_\theta = M' \mid_\sigma$. Then $M'_1 := M' \underset{M'\mid_\sigma}{+} M_1 \in \textbf{Mod}(T'_1)$ is a witness for $M' \models_{T'} \text{derive from } T_1 \text{ by } T' \xrightarrow{\theta'} T_1$.

Having amalgamation is a property which $\textbf{derive!}(\mathcal{F})$ immediately inherits from $\mathcal{F}$.

Now $\textbf{derive!}(\Phi, \beta) := (\Phi, \Phi, \beta)$ as in Proposition 4.3.2. The crucial point to show is that for $\text{derive from } T_1 \text{ by } T \xrightarrow{\theta} T_1 \in sen^{\textbf{derive!}(\mathcal{F})}(T)$, derive from $\Phi(T_1)$ by $\Phi(T) \xrightarrow{\Phi(\theta)} \Phi(T_1) \in sen^{\textbf{derive!}(\mathcal{F})}(\Phi(T))$. Let $M_1', M_2' \in$ $\textbf{Mod}'(\Phi(T_1))$ with $M_1'|_{\Phi(\theta)} = M_2'|_{\Phi(\theta)}$.

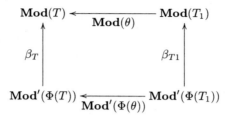

This implies $\beta_{T_1}(M_1')|_\theta = \beta_T(M_1'|_{\Phi(\theta)}) = \beta_T(M_2'|_{\Phi(\theta)}) = \beta_{T_1}(M_2')|_\theta$. Since **derive from** T_1 **by** $T \xrightarrow{\theta} T_1$ is a T-sentence in $\textbf{derive!}(\mathcal{F})$, we get $\beta_{T_1}(M_1') = \beta_{T_1}(M_2')$. Since the above diagram allows amalgamation, both M_1' and M_2' are amalgamations $M_1'|_{\Phi(\theta)} \underset{\beta_T(M_1'|_{\Phi(\theta)})}{+} \beta_{T_1}(M_1')$. By uniqueness of amalgamation, we get $M_1' = M_2'$.

Preservation of amalgamation carries over directly from (Φ, β) to (Φ, Φ, β).

We next have to show that $\eta_\mathcal{I}$ from Proposition 4.3.2 is already in $\textbf{Inst}^{amal}$, and that ϵ_F is in $SpecFram^{amal}$.

The former follows from Proposition 2.3.7. Concerning the latter, to be able to apply Proposition 2.3.7 to $\epsilon_\mathcal{F} = (\Phi, \beta)$, we have to show that β is an isomorphism. Recall that for a presentation $TT = (T, \Gamma)$ with $(T_0, \rho: T \longrightarrow T_0, (\rho_\theta: T_\theta \longrightarrow T_0)_{\textbf{derive from } T_\theta \textbf{ by } T \xrightarrow{\theta} T_\theta \in \Gamma}) = Colim(\mathcal{T})$ with $\mathcal{T} = (T, (T \xrightarrow{\theta} T_\theta)_{\textbf{derive from } T_\theta \textbf{ by } T \xrightarrow{\theta} T_\theta \in \Gamma})$, we have defined β_{TT} to be $\textbf{Mod}(\rho)$. Its inverse is β_{TT}^{-1} which takes a TT-model MM to the amalgamated sum $\underset{MM}{+}(M_\theta)_{\textbf{derive from } T_\theta \textbf{ by } \theta \in \Gamma}$, where M_θ is the unique θ-expansion of MM. Then $\beta_{TT}(\beta_{TT}^{-1}(MM)) = (\underset{MM}{+}(M_\theta)_{\textbf{derive from } T_\theta \textbf{ by } \theta \in \Gamma})|_\rho = MM$, and for a T'-model M', we have that $\beta_{TT}^{-1}(\beta_{TT}(M')) = \underset{M'|_\rho}{+}(M_\theta)_{\textbf{derive from } T_\theta \textbf{ by } \theta \in \Gamma}$ where M_θ is the unique θ-expansion of $M'|_\rho$. Since $M'|_{\rho_\theta}$ is another such expansion, $M'|_{\rho_\theta} = M_\theta$, and, by the uniqueness of the amalgamation, $\underset{M'|_\rho}{+}(M_\theta)_{\textbf{derive from } T_\theta \textbf{ by } \theta \in \Gamma} = M'$.

The adjointness equations follow as in Proposition 4.3.2. □

Again, the adjunction induces a monad $\textbf{Ext} = (\textbf{derive!} \circ \textbf{Pres}, \eta, \textbf{derive!}(\epsilon_{\textbf{Pres}}))$ on $\textbf{Inst}^{amal}$ with Kleisli category $\textbf{Inst}^{amal}_{\textbf{Ext}}$.

4.3. THE PRESENTATION EXTENSION MONAD AND WEAK REPRESENTATIONS OF INSTITUTIONS

Proposition 4.3.5 The functor $K: \mathbf{Inst}^{amal}_{\mathbf{Ext}} \longrightarrow \mathbf{WeakInst}^{amal}$ being the identity on objects and $K_{\mathbf{Ext}}$ on morphisms is an isomorphism.

Proof:

The comparison functor $K_{\mathbf{Ext}}$ is full and faithful (see A.1). Spelled out in more detail, $K(\mathcal{I} \xrightarrow{\mu} \mathbf{derive}!(\mathbf{Pres}(\mathcal{I})))$ is $\mathbf{Pres}(\mathcal{I}) \xrightarrow{\mathbf{Pres}(\mu)} \mathbf{Pres}(\mathbf{derive}!(\mathbf{Pres}(\mathcal{I}'))) \xrightarrow{\epsilon_{\mathbf{Pres}(\mathcal{I}')}} \mathbf{Pres}(\mathcal{I}')$, while $K^{-1}(\mathbf{Pres}(\mathcal{I}) \xrightarrow{\mu} \mathbf{Pres}(\mathcal{I}'))$ is $\mathcal{I} \xrightarrow{\eta_{\mathcal{I}}} \mathbf{derive}!(\mathbf{Pres}(\mathcal{I})) \xrightarrow{\mathbf{derive}!(\mu)} \mathbf{derive}!(\mathbf{Pres}(\mathcal{I}'))$. If $\mu = (\Phi, \alpha, \beta)$, the latter maps a signature Σ to $\Phi(\Sigma, \emptyset)$, a Σ-sentence φ to **derive from** $\Phi(\Sigma, \{\varphi\})$ **by** $\Phi(\Sigma, \emptyset) \hookrightarrow \Phi(\Sigma, \{\varphi\})$ and a $((\Sigma, \Gamma), \Gamma')$-model M to $\beta_{(\Sigma, \Gamma)}(M)$. □

Thus a weak institution representation $\mu: \mathbf{Pres}(\mathcal{I}) \longrightarrow \mathbf{Pres}(\mathcal{I}') \in \mathbf{WeakInst}^{amal}$ mapping presentations to presentations and models to models can be modularized as a simple institution representation $K^{-1}(\mu): \mathcal{I} \longrightarrow \mathbf{derive}!(\mathbf{Pres}(\mathcal{I}'))$ mapping signatures to presentations and mapping sentences to presentation extensions (while the mapping of models is kept). Note that μ applied to a presentation just is the multiple pushout of all presentation extensions resulting from the application of $K^{-1}(\mu)$ to each sentence of the presentation.

Further we have the same diagram as in section A.1:

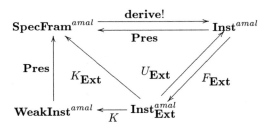

$F_{\mathbf{Ext}}$ is just the inclusion: every simple representation is also a weak representation. $U_{\mathbf{Ext}}$ considers a weak representation in Kleisli form $\mu: \mathcal{I} \longrightarrow \mathbf{derive}!(\mathbf{Pres}(\mathcal{I}))$ to live in $\mathbf{derive}!(\mathbf{Pres}(\mathcal{I}))$ — an institution where all presentation extensions are available as sentences.

It is easy to see that each conjunctive representation also is a weak institution representation, this gives us an embedding $weak_\wedge: \bigwedge \mathbf{Inst}^{amal} \longrightarrow \mathbf{WeakInst}$.

By the way: by Theorem 6.1.2 and Corollary 5.1.2, there is no weak representation from limit theories $R(R \Longrightarrow \exists! R =)$ to Horn Clause theories $R(R \Longrightarrow R =)$, that is, the meta-level **derive!** cannot simulate the object-level $\exists!$ in this case.

We also can use the adjunction between **SpecFram**^{amal} and **Inst**^{amal} for borrowing logical structure. A specification frame $\mathcal{F} = (\mathbf{Th}, \mathbf{Mod})$ may borrow sentences from an institution $\mathcal{I}'$ along a specification frame representation $\mu = (\Phi, \beta): \mathcal{F} \longrightarrow \mathbf{Pres}(\mathcal{I}')$. Given such a representation, let $\mathcal{I}$ be the institution $(\mathbf{Th}^{\mathcal{F}}, sen, \mathbf{Mod}^{\mathcal{F}}, \models)$ with $sen(T) = \{(T \xrightarrow{\sigma} T', \varphi) \mid \Phi(\sigma) = \text{derive from } (\Sigma, \Gamma \cup \{\varphi\}) \text{ by } (\Sigma, \Gamma) \longrightarrow (\Sigma, \Gamma \cup \{\varphi\}) \in sen^{\mathbf{derive!}(\mathbf{Th_0}(\mathcal{I}'))}(\Sigma)\}$ and $M \models_{\Sigma} (T \xrightarrow{\sigma} T', \varphi)$ iff $M \models_{\Sigma}^{\mathbf{derive!}(\mathcal{F})}$ **derive from** T' **by** $T \xrightarrow{\sigma} T'$.

Theorem 4.3.6 Let $\Psi(T) = (T, \emptyset)$. The following diagram is a pullback in **Inst**:

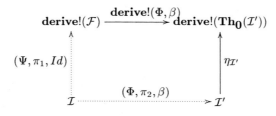

Proof:
This follows from a generalization of the construction of limits in **PlainInst** in [Tar96b] to **Inst**. □

That is, $\mathcal{F}$ can borrow those sentences from $\mathcal{I}'$ which do not lead to new presentations. For example, if we take $\mathcal{F}$ to be Lawvere theories or FP-sketches over **Set** [BW85] (which are mathematically elegant formulations of equational theories but lack a notion of axiom) and map them to their equational theories (therefore we have to equip equational theories with some notion of derived signature morphism), we can use usual equations as new axioms.

4.4 Semantical consequence in derive!($\mathcal{F}$) and derive!(Pres($\mathcal{I}$))

For weak institution representations, there is no theorem like Corollary 4.2.2 that allows us to re-use theorem provers. But at least, since a weak institution representation $\mu: \mathbf{Pres}(\mathcal{I}) \longrightarrow \mathbf{Pres}(\mathcal{I}')$ can be modularized as a simple institution representation $K^{-1}(\mu): \mathcal{I} \longrightarrow \mathbf{derive!}(\mathbf{Pres}(\mathcal{I}'))$, we know that if model translation is surjective, then a theorem prover for **derive!**($\mathbf{Pres}(\mathcal{I}')$) can be

4.4. SEMANTICAL CONSEQUENCE IN derive!($\mathcal{F}$) AND derive!(Pres($\mathcal{I}$))

re-used also for $\mathcal{I}$. But what does semantical consequence in **derive!(Pres($\mathcal{I}'$))** look like? To be able to answer this question, we first need

Definition 4.4.1 Given a specification frame $\mathcal{F} = (\mathbf{Th}, \mathbf{Mod})$, a presentation morphism $\sigma: T \longrightarrow T'$ in **Th** is called *strongly persistent*, if $\mathbf{Mod}(\sigma)$ has a left adjoint with unit being the identity.

For arbitrary sentences, we only have the following:

Proposition 4.4.2 Given a specification frame $\mathcal{F}$ with amalgamation, and a presentation $TT = (T, \Gamma)$ in **derive!($\mathcal{F}$)**, let $(T_0, \rho: T \longrightarrow T_0, (\rho_\theta: T_\theta \longrightarrow T_0)_{T \xrightarrow{\theta} T_\theta \in \Gamma}) = Colim(\mathcal{T})$ where $\mathcal{T} = (T, (T \xrightarrow{\theta} T_\theta))_{\text{derive from } T_\theta \text{ by } T \xrightarrow{\theta} T_\theta \in \Gamma})$. For a sentence **derive from** T_1 **by** $T \xrightarrow{\sigma} T_1$ in **derive!($\mathcal{F}$)**, let T_2 be constructed by the pushout

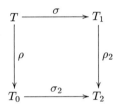

Then
$$TT \models_T T \xrightarrow{\sigma} T_1$$

if and only if there is a function F from $|\mathbf{Mod}(T_0)|$ to $|\mathbf{Mod}(T_2)|$ such that for all $M \in |\mathbf{Mod}(T_0)|$, $F(M)|_{\sigma_2} = M$. In particular this is the case if σ_2 is strongly persistent.

Proof:

$TT \models_T T \xrightarrow{\sigma} T_1$ iff each TT-model is also a σ-model iff for each T_0-model M, $M|_\rho$ has a σ-expansion iff there is a function from T_0-models to T_1-models yielding for $M \in |\mathbf{Mod}(T_0)|$ a σ-expansion of $M|_\rho$ iff (by amalgamation) there is a function from T_0-models to T_2-models yielding for $M \in |\mathbf{Mod}(T_0)|$ a σ_2-expansion of M.

In the case of strong persistency of σ_2, F^{σ_2} yields this σ_2-expansion. □

Proposition 4.4.3 Given an institution $\mathcal{I}$ with amalgamation and a presentation
$$TT = ((\Sigma, \Gamma), \Gamma\Gamma)$$

in **derive!(Pres($\mathcal{I}$))** let $((\Sigma_0, \Gamma_0), \rho: (\Sigma, \Gamma) \longrightarrow (\Sigma_0, \Gamma_0), (\rho_\theta: (\Sigma_\theta, \Gamma_\theta) \longrightarrow (\Sigma_0, \Gamma_0))_{\text{derive from } T_\theta \text{ by } \theta \in \Gamma\Gamma})$ be the colimit as above. For a sentence $\sigma =$

$(\Sigma, \Gamma) \hookrightarrow (\Sigma, \Gamma \cup \Gamma')$, let (Σ_2, Γ_2) be constructed by a pushout as above. Then

$$TT \models_{(\Sigma,\Gamma)} \textbf{derive from } (\Sigma, \Gamma \cup \Gamma') \textbf{ by } (\Sigma, \Gamma) \hookrightarrow (\Sigma, \Gamma \cup \Gamma')$$

if and only if

$$\Gamma_0 \models_{\Sigma_0} \rho[\Gamma']$$

in $\mathcal{I}$.

Proof:
$TT \models_{(\Sigma,\Gamma)} \textbf{derive from } (\Sigma, \Gamma \cup \Gamma') \textbf{ by } (\Sigma, \Gamma) \hookrightarrow (\Sigma, \Gamma \cup \Gamma')$ iff each TT-model is also a σ-model iff for each (Σ_0, Γ_0)-model M, $M \mid_\rho$ has a σ-expansion iff for each (Σ_0, Γ_0)-model M, $M \mid_\rho \models_\Sigma \Gamma'$ iff for each (Σ_0, Γ_0)-model M, $M \models_{\Sigma_0} \rho[\Gamma']$ iff $\Gamma_0 \models_{\Sigma_0} \rho[\Gamma']$. □

Thus, given a weak representation $\mu: \mathcal{I} \longrightarrow \mathcal{I}'$ with surjective model translation, if we have a theorem prover for $\mathcal{I}'$ and a checker for strong persistency of presentation extensions in $\mathcal{I}'$, by the above propositions we get a sound theorem prover for $\mathcal{I}$ which is guaranteed to be complete only for those sentences which are translated by the modularized weak representation to a presentation extension which is the identity on signatures.

4.5 The model class monad and semi-representations of institutions

In the previous section, we generated sentences from the presentation extensions of a specification frame. Now we explore the other canonical possibility: to generate sentences from model classes.

Proposition 4.5.1 For the functor $frame: \textbf{PlainInst} \longrightarrow \textbf{SpecFram}$, there exists a right adjoint $(\_)^\heartsuit: \textbf{SpecFram} \longrightarrow \textbf{PlainInst}$.

Proof:
Let $(\textbf{Th}, \textbf{Mod})^\heartsuit = (\textbf{Th}, sen, \textbf{Mod}, \models)$ with $sen(T) = \{ \mathcal{M} \mid \mathcal{M} \subseteq |\textbf{Mod}(T)| \}^2$ and $sen(T \xrightarrow{\sigma} T')(\mathcal{M}) = (\textbf{Mod}(\sigma))^{-1}(\mathcal{M})$. Satisfaction is just the element relation. $(\Phi, \beta)^\heartsuit = (\Phi, \beta^{-1}, \beta)$.

The satisfaction condition is easy: $M \mid_\sigma \models \mathcal{M}$ iff $M \mid_\sigma \in \mathcal{M}$ iff $M \in \textbf{Mod}(\sigma)^{-1}[\mathcal{M}]$ iff $M \in sen(\sigma)(\mathcal{M})$ iff $M \models sen(\sigma)(\mathcal{M})$.

---
[2]Strictly speaking, here we have to change the definition of institution to allow also proper classes of sentences. The foundational problem with that is the same as that with proper classes of models, which are there anyway. See section A.5.

4.5. THE MODEL CLASS MONAD AND SEMI-REPRESENTATIONS OF INSTITUTIONS

The unit of the adjunction $\eta_\mathcal{I}: \mathcal{I} \longrightarrow (frame(\mathcal{I}))^\heartsuit$ is given by $\eta_\mathcal{I} = (id, \overline{\alpha}, id)$ with $\overline{\alpha}_\Sigma(\varphi) = \{ M \in \mathbf{Mod}(\Sigma) \mid M \models_\Sigma \varphi \}$. The counit $\epsilon_\mathcal{F}: frame(\mathcal{F}^\heartsuit) \longrightarrow \mathcal{F}$ is just the identity.

To show adjointness, we must prove that

$$\mathcal{F}^\heartsuit \xrightarrow{\eta_{(\mathcal{F})^\heartsuit}} (frame(\mathcal{F}^\heartsuit))^\heartsuit \xrightarrow{(\epsilon_\mathcal{F})^\heartsuit} \mathcal{F}^\heartsuit = id_{\mathcal{F}^\heartsuit}$$

and

$$frame(\mathcal{I}) \xrightarrow{frame(\eta_\mathcal{I})} frame(frame(\mathcal{I})^\heartsuit) \xrightarrow{\epsilon_{frame(\mathcal{I})}} frame(\mathcal{I}) = id_{frame(\mathcal{I})}$$

The former follows by the remark that $\mathcal{F}^\heartsuit$ already has model classes as sentences, so $\overline{\alpha}_\Sigma$ is just the identity on sentences. The latter follows from $frame(\eta_\mathcal{I}) = frame(id, \overline{\alpha}, id) = (id, id)$. □

The adjoint situation $(\eta, \epsilon): frame \dashv (\_)^\heartsuit: \mathbf{SpecFram} \longrightarrow \mathbf{PlainInst}$ given above induces a monad $\mathbf{Mod}$ on $\mathbf{PlainInst}$.

Definition 4.5.2 The morphisms of the Kleisli category $\mathbf{SemiInst} = \mathbf{PlainInst_{Mod}}$ are called *institution semi-representations*.

Since is full and faithful, semi-representations $\mu: \mathcal{I} \longrightarrow \mathcal{I}'$ are (in one-one correspondence with) specification frame representations $K_{\mathbf{Mod}}(\mu): frame(\mathcal{I}) \longrightarrow frame(\mathcal{I}')$, that is, just a plain representation without a component mapping sentences. The correspondence is given by $K_{\mathbf{Mod}}$ mapping $\mu: \mathcal{I} \longrightarrow (frame(\mathcal{I}'))^\heartsuit$ to $frame(\mathcal{I}) \xrightarrow{frame(\mu)} frame((frame(\mathcal{I}'))^\heartsuit) \xrightarrow{id_{frame(\mathcal{I}')}} frame(\mathcal{I}')$ and $K_{\mathbf{Mod}}^{-1}$ mapping $\mu': frame(\mathcal{I}) \longrightarrow frame(\mathcal{I}')$ to $\mathcal{I} \xrightarrow{\eta_\mathcal{I}} (frame(\mathcal{I}))^\heartsuit \xrightarrow{\mu'^\heartsuit} (frame(\mathcal{I}'))^\heartsuit$. Such semi-representations are related to institution semi-morphisms [Tar96a], which are defined similarly except that signatures and models are mapped covariantly. For an example, see [Tar96a].

I do not know if there are good examples of institution semi-representations that are not already institution representations. But at least they might be used to formalize some concepts in finite model theory. For example, let $\mu = (\Phi, \beta)$ be the institution semi-representation from the institution of unsorted first-order logic plus existential second-order quantifiers to ($\mathbf{Th_0}$ applied to) the institution of unsorted first-order logic plus least fixed-point operators for formulas expressing relations, and model categories restricted to finite models. Φ maps a signature Σ to Σ plus a binary relation symbol $\leq$ and axioms stating $\leq$ to be a total ordering. β just forgets the total ordering of a $\Phi(\Sigma)$-model.

Then, by the results of Fagin and Immerman [Lei87], μ can be extended to a simple institution representation iff P=NP.

Again, we have the possibility of borrowing sentences from an institution along a specification frame representation $\mu = (\Phi, \beta): \mathcal{F} \longrightarrow frame(\mathcal{I}')$. Given such a representation, let $\mathcal{F} = (\mathbf{Th}, \mathbf{Mod})$. We put $\mathcal{I} := (\mathbf{Th}, sen, \mathbf{Mod}, \models)$, where $sen(T) =$
$$\{ (\mathcal{M}, \varphi) \mid \mathcal{M} \subseteq \mathbf{Mod}(T), \varphi \in sen'(\Phi(T)) \text{ and } \beta_T^{-1}[\mathcal{M}] = \mathbf{Mod}(\Phi(T), \{\varphi\}) \}$$
and $M \models_T (\mathcal{M}, \varphi)$ iff $M \in \mathcal{M}$.

Theorem 4.5.3 The following diagram is a pullback

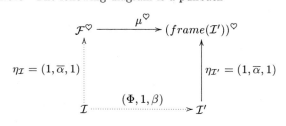

Proof:
Again, this follows from the construction of limits in **PlainInst** in [Tar96b]. □

Thus, in comparison to theorem 4.3.6, here we can borrow all sentences from an institution which capture some β-preimage of some class of models in $\mathcal{F}$. If β is the identity (or an isomorphism), this means we can use all $\mathcal{I}'$-sentences which are semantically captured by some theory in $\mathcal{F}$.

Since the specification frame $frame(\mathcal{I})$ is very poor, in applications, it may be more useful to consider $frame(\mathbf{Pres}(\mathcal{I}))$. However, the composite of $frame$ and **Pres** probably is not adjoint (note that $frame$ is left adjoint, while **Pres: Inst** $\longrightarrow$ **PlainInst** is right adjoint).

For example, if $\mathcal{F}$ are Lawvere theories or sketches, we now can borrow also conditional equations and thus strictly enlarging the power of $\mathcal{F}$.

4.6 Summary

We have started with the argument that there are problems with the distinction between signatures and sentences within the notion of institution. While the distinction is useful and necessary, representations between institutions often have to go beyond this distinction. Therefore, three new notions of representation between institutions, conjunctive representations, weak representations

4.6. Summary

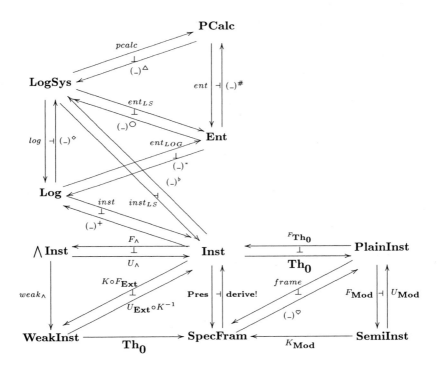

Figure 4.1: Functors and adjunctions between different categories of logical systems

and semi-representations, have been introduced in this chapter. These allow to relax the distinction between signatures and sentences while still being sentence-structured, opposed to mere representations of specification frames.

When introducing these types of arrow we have followed some guidelines that allow to relate different categories of logical system by a pair of adjoint functors. Along two of those pairs, there is a borrowing of sentences from an institution, extending the work of Cerioli and Meseguer on borrowing logical structure along representations [CM93] to the relation between specification frames and institutions.

The technical results of the chapter are summarized in the lower half of Fig. 4.1, where the arrows **derive!**, $U_{\mathbf{Ext}} \circ K^{-1}$ and $K \circ F_{\mathbf{Ext}}$ only exist when everything is restricted either to the case with composable signatures (and **Inst** replaced by **Instrps**) or to the case with amalgamation.

4.7 Subinstitutions and equivalent expressiveness of institutions

What does equivalent expressiveness of logical systems mean? In this section, I develop an answer to this question. Therefore, among the institution representations, I single out a subclass called embeddings, which represent faithfully all essential properties. With this notion of embedding, later on, separation results between institutions can be proved.

Meseguer [Mes89] defines a notion of subinstitution. His definition is as follows:

Definition 4.7.1 (Meseguer) A *subinstitution* is an institution representation $\mu = (\Phi, \alpha, \beta): \mathcal{I} \longrightarrow \mathcal{I}'$ with Φ faithful and injective on objects (that is, Φ is an embedding of categories), with α pointwise injective and with β a natural isomorphism.

Now this definition is somewhat too narrow, because there are many situations where β is just a pointwise equivalence of categories. The most prominent example is the embedding of order-sorted algebra into many-sorted algebra, see [GM92] and theorem 5.1.1, (5.14). Now equivalences of categories still preserve and reflect all interesting categorical properties. So we restate the definition of subinstitution as follows:

Definition 4.7.2 An *embedding of institutions* is an institution representation $\mu = (\Phi, \alpha, \beta): \mathcal{I} \longrightarrow \mathcal{I}'$, with Φ an embedding, with α pointwise injective and with β a pointwise equivalence of categories. If there is such an embedding, $\mathcal{I}$ is called *subinstitution* of $\mathcal{I}'$. □

In [KM95], it was additionally required that Φ preserves colimits, in order to faithfully represent Burstall's and Goguen's slogan "putting theories together to make specifications" [BG77]. But this is reasonable only when signatures are composable, and thus should be required only in this situation (see section 4.3).

Definition 4.7.3 We call two institutions *equivalent in expressiveness*, if there are embeddings of institutions in both directions. □

This implies that satisfaction is represented faithfully, and we can prove a property similar to the defining property of Salibra's and Scollo's invertible pre-institution transformations [SS92]:

4.7. Subinstitutions and Equivalent Expressiveness of Institutions

Proposition 4.7.4 Let $\mathcal{I} = (\textbf{Sign}, sen, \textbf{Mod}, \models)$ and $\mathcal{I}' = (\textbf{Sign}', sen', \textbf{Mod}', \models')$ be two institutions in which model isomorphisms preserve satisfaction and let $\mathcal{I}$ and $\mathcal{I}'$ be equivalent in expressiveness via embeddings $\mu^1 = (\Phi^1, \alpha^1, \beta^1): \mathcal{I} \longrightarrow \mathcal{I}'$ and $\mu^2 = (\Phi^2, \alpha^2, \beta^2): \mathcal{I}' \longrightarrow \mathcal{I}$. Let $\beta^1{}'_\Sigma$ be adjoint to β^1_Σ for $\Sigma \in \textbf{Sign}$ and $\beta^2{}'_{\Sigma'}$ be adjoint to $\beta^2_{\Sigma'}$ for $\Sigma' \in \textbf{Sign}'$. Then

$$M \models_\Sigma \varphi \iff (\beta^2{}'_{\Phi(\Sigma)})(\beta^1{}'_\Sigma)(M) \models_{\Phi'\Phi(\Sigma)} \alpha'_{\Phi\Sigma} \alpha_\Sigma(\varphi)$$

Proof: Apply the satisfaction condition and the condition that model isomorphisms preserve satisfaction each twice. □

Following Salibra and Scollo, two pre-institutions have equivalent expressiveness if and only if there is an invertible pre-institution transformation between them. Our notion of equivalence in expressiveness is much stronger, since it considers more than logical consequence: If we have two institutions that are equivalent in expressiveness, each sentence, theory, model category, theory morphism, colimit of theories, reduct etc., taken in one institution, can be faithfully simulated in the other. With this, it is easy to show that constructions not only of Cerioli's *representation independent* specification language constructs [Cer93], but of any reasonable[3] specification language constructs that can be formulated in an institution independent way [ST88a] are preserved in both directions.

From the notion of embedding of institutions, other notions of embedding can now be derived:

Definition 4.7.5 An *embedding of specification frames* is a specification frame representation $\mu = (\Phi, \beta): \mathcal{F} \longrightarrow \mathcal{F}'$, with Φ an embedding and with β a pointwise equivalence of categories. If there is such an embedding, $\mathcal{F}$ is called *subframe* of $\mathcal{F}'$. □

Definition 4.7.6 A *simple (conjunctive, weak)* embedding of institutions is a simple (conjunctive, weak) institution representation $\mu: \mathcal{I} \longrightarrow T(\mathcal{I}')$ (where T is the corresponding monadic construction, see chapter 4) with μ being an embedding, considered as a (plain) institution representation (resp. specification frame representation). If there is such an embedding, $\mathcal{I}$ is called *simple (conjunctive, weak) subinstitution* of $\mathcal{I}'$.

Definition 4.7.7 We call two institutions *simply (conjunctively, weakly) equivalent in expressiveness*, if there are simple (conjunctive, weak) embeddings of institutions in both directions. □

---

[3]i.e. not using tricks like that in the proof of proposition 4.8.2 below.

Proposition 4.7.8 Suppose that there is a specification frame representation $(\Phi, \beta): \mathcal{F} \longrightarrow \mathcal{F}'$ in **SpecFram**^{comp}. If β_T is an isomorphism of categories for each theory T, then

$$\mathcal{F}' \text{ has amalgamation} \Rightarrow \mathcal{F} \text{ has amalgamation}$$

Proof:

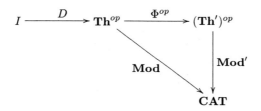

We have $\mathbf{Mod} \cong \mathbf{Mod}' \circ \Phi^{op}$ by β. Let $D: I \longrightarrow \mathbf{Th}$ be a source considered as a diagram. Then $\mathbf{Mod}(Lim(D)) \cong \mathbf{Mod}'(\Phi^{op}(Lim(D)))$. Since Φ^{op} preserves multiple pullbacks, this is equal to $\mathbf{Mod}'(Lim(\Phi^{op} \circ D))$. Since $\mathbf{Mod}'$ preserves multiple pullbacks, this in turn is equal to $Lim(\mathbf{Mod}' \circ \Phi^{op} \circ D)$. Since $\mathbf{Mod} \cong \mathbf{Mod}' \circ \Phi^{op}$ and Lim is functorial (see [Lan72]), this is isomorphic to $Lim(\mathbf{Mod} \circ D)$. Since limits are translated by isomorphisms to limits, $\mathbf{Mod}$ preserves multiple pullbacks. □

Proposition 4.7.9 Suppose that there is a specification frame representation $(\Phi, \beta): \mathcal{F} \longrightarrow \mathcal{F}'$. If β_T is an equivalence of categories for each theory T (in particular, if (Φ, β) is an embedding), then

$$\mathcal{F}' \text{ is liberal} \Rightarrow \mathcal{F} \text{ is liberal}$$

Proof:

Let $\sigma: T \longrightarrow T'$ be a theory morphism is $\mathcal{F}$. Since β_T and $\beta_{T'}$ are equivalences of categories, we can choose β_T' left adjoint to β_T and $\beta_{T'}'$ right adjoint to $\beta_{T'}$ such that units and counits are isomorphisms. By naturality of β, $\beta_T \circ \mathbf{Mod}^{\mathcal{F}'}(\Phi(\sigma)) = \mathbf{Mod}^{\mathcal{F}}(\sigma) \circ \beta_{T'}$. Multiplying right with $\beta_{T'}'$ yields $\beta_T \circ \mathbf{Mod}^{\mathcal{F}'}(\Phi(\sigma)) \circ \beta_{T'}' \cong \mathbf{Mod}^{\mathcal{F}}(\sigma)$. By liberality of $\mathcal{F}'$, $\mathbf{Mod}^{\mathcal{F}'}(\Phi(\sigma))$ has a left adjoint $F_{\Phi(\sigma)}$. By compositionality of left adjoints, $\beta_{T'} \circ F_{\Phi(\sigma)} \circ \beta_T'$ is left adjoint to $\mathbf{Mod}^{\mathcal{F}}(\sigma)$. □

4.8 Is our notion of equivalence strong enough? An example

As an example for testing the notion of embedding, we use the institutions $R(R \Longrightarrow R =)$ and $P(\stackrel{f}{=}\Rightarrow\stackrel{f}{=})$ (see section 3.3), which can be embedded into each other in the following way: simply represent relations by domains of partial operations and partial operations by their graph relations. This can be formalized as follows:

Proposition 4.8.1 $R(R \Longrightarrow R =)$ and $P(\stackrel{f}{=}\Rightarrow\stackrel{f}{=})$ are simply equivalent in expressiveness.

Proof: For the first direction, $R(R \Longrightarrow R =)$ is a simple subinstitution of $P(\stackrel{f}{=}\Rightarrow\stackrel{f}{=})$. Define $\mu^{chardom} = (\Phi^{chardom}, \alpha^{chardom}, \beta^{chardom}) : R(R \Longrightarrow R =) \longrightarrow P(\stackrel{f}{=}\Rightarrow\stackrel{f}{=})$ by

- $\Phi^{chardom}(S, OP, REL)$ is the theory with signature

  $(S \cup \{\,one\,\},$
  $OP \cup \{\,*: one\,\},$
  $\{\chi^R : s_1, \ldots, s_n \longrightarrow one \mid R : s_1, \ldots, s_n \in REL\,\})$

  and the axiom $\forall x, y : one \,.\, x = y$. This is extended to signature morphisms in the obvious way.

- $\alpha^{chardom}_\Sigma$ doesn't change the variable system. On atomic formulas, the action is

  $\alpha^{chardom}_\Sigma [\![t_1 = t_2]\!] = [\![t_1 = t_2]\!]$

  $\alpha^{chardom}_\Sigma [\![R(t_1, \ldots, t_n)]\!] = [\![\chi^R(t_1, \ldots, t_n) \stackrel{f}{=} *]\!]$

  This can be easily extended to conditional formulas and is compatible with signature morphisms (that is, α is natural).

- $\beta^{chardom}_{(S,OP,REL)}(A') = ((A'_s)_{s \in S}, (op_{A'})_{op \in OP}, (\text{dom } \chi^R_{A'})_{R \in REL})$, and
  $\beta^{chardom}_{(S,OP,REL)}(h') = h'$

  Because of the axioms in $\Phi^{chardom}(\Sigma)$, the partial operations of a $\Phi^{chardom}(\Sigma)$-model A' are exactly determined by their domains. Therefore, β can be see to be an isomorphism: Its inverse adds a one-point carrier for sort one and replaces a relation R_A of a Σ-model A by the partial function with domain R_A (note that partial functions to the one-point carrier are exactly determined by their domain).
  Now, taking reducts is orthogonal to the above interchange between relations and partial functions. This gives us the naturality of $\beta^{chardom}$.

The representation condition can be verified as follows: Since $R_{\beta_T^{chardom}(A')} =$ dom $\chi_{A'}^R$, a valuation $\nu\colon X \longrightarrow A'$ satisfies the atomic formula $\chi^R(t_1,\ldots,t_n) \stackrel{f}{=} *$ if and only if ν, considered as a valuation $\nu\colon X \longrightarrow \beta_T^{chardom}(A')$, satisfies $R(t_1,\ldots,t_n)$. This can be extended to conditional formulas, as usual.

Now the converse direction: $P(\stackrel{f}{=}\Rightarrow\stackrel{f}{=})$ is a simple subinstitution of $R(R \Longrightarrow R=)$: Define $\mu^{graph} = (\Phi^{graph},\alpha^{graph},\beta^{graph})\colon P(\stackrel{f}{=}\Rightarrow\stackrel{f}{=}) \longrightarrow R(R \Longrightarrow R=)$ by

- $\Phi^{graph}(S,OP,POP)$ is the theory with signature

$$(S, OP, \{\, G^{pop} : s_1,\ldots,s_n\, s \mid pop\colon s_1,\ldots,s_n \longrightarrow s \in POP\,\})$$

  and axioms

$$\forall x_1\colon s_1,\ldots,x_n\colon s_n, y\colon s, z\colon s\,.\\
G^{pop}(x_1,\ldots,x_n,y) \wedge G^{pop}(x_1,\ldots,x_n,z) \Longrightarrow y = z$$

  for each $pop \in POP$. This is extended to signature morphisms in the obvious way.

- α_Σ^{graph} doesn't change the variable system. On atomic formulas, the action is

$$\alpha_\Sigma^{graph}\llbracket t_1 = t_2 \rrbracket = \llbracket t_1 = t_2 \rrbracket$$

$$\alpha_\Sigma^{graph}\llbracket pop(t_1,\ldots,t_n) \stackrel{f}{=} t \rrbracket = \llbracket G^{pop}(t_1,\ldots,t_n,t)\rrbracket$$

- $\beta_{((S,OP,POP),\Gamma)}^{graph}(A') = ((A'_s)_{s\in S},(op_{A'})_{op\in OP},(\text{graph}^{-1}(G_{A'}^{pop}))_{pop\in POP})$, where graph^{-1} takes the graph of a partial operation to that operation, and $\beta_{((S,OP,POP),\Gamma)}^{graph}(h') = h'$. The $\Phi^{graph}(\Sigma)$-axioms guarantee that $G_{A'}^{op}$ actually is the graph of a partial operation. Moreover, β^{graph} is an isomorphism: its inverse replaces the partial operations of a model by their graphs.

The representation condition is fulfilled, because a valuation $\nu\colon X \longrightarrow A'$ (also considered as $\nu\colon X \longrightarrow \beta_T^{graph}(A')$) satisfies the atomic formula $G^{op}(t_1,\ldots,t_n,t)$ if and only if $(\nu^\#(t_1),\ldots,\nu^\#(t_n),\nu^\#(t)) \in G_{A'}^{op}$ if and only if graph$^{-1}(G_{A'}^{op})(\nu^\#(t_1),\ldots,\nu^\#(t_n))$ is defined and equal to $\nu^\#(t)$ if and only if $\nu\colon X \longrightarrow \beta_T^{graph}(A')$ satisfies $op(t_1,\ldots,t_n) \stackrel{e}{=} t$.

That Φ preserves multiple pushouts (even all colimits) follows from left adjointness of Φ in both cases by Proposition A.2.3. The right adjoint $\Phi^{graph'}$ to Φ^{graph} maps a theory $((S,OP,REL),\Gamma)$ to the signature

$$(S, OP, \{\, pop\colon s_1,\ldots,s_n \longrightarrow s \mid G^{pop} : s_1,\ldots,s_n\, s \in REL \text{ and}\\
\forall x_1\colon s_1,\ldots,x_n\colon s_n, y\colon s, z\colon s\,.\\
G^{pop}(x_1,\ldots,x_n,y) \wedge G^{pop}(x_1,\ldots,x_n,z) \Longrightarrow y = z \in \Gamma\,\})$$

4.8. Is our notion of equivalence strong enough? An example

The unit is the identity. To prove the universal property, note that a signature morphism $\sigma\colon (S, OP, POP) \longrightarrow \Phi^{graph'}((S, OP, REL), \Gamma)$ can be seen as theory morphism $\sigma\colon \Phi^{graph}(S, OP, POP) \longrightarrow ((S, OP, REL), \Gamma)$. For $\Phi^{chardom}$, the proof of left-adjointness is similar.

Now, $\mu^{chardom}$ and μ^{graph} show that $R(R \Longrightarrow R =)$ and $P(\stackrel{f}{=}\Longrightarrow\stackrel{f}{=})$ are simple subinstitutions of each other, therefore, $R(R \Longrightarrow R =)$ and $P(\stackrel{f}{=}\Longrightarrow\stackrel{f}{=})$ are simply equivalent in expressiveness. □

$\Phi^{chardom}$ and Φ^{graph} are left-adjoint and can be shown to be full. The only thing missing them to be a categorical equivalence is isomorphism-denseness (see definition A.2.1). $\Phi^{flatten}$ in chapter 7 is isomorphism-dense but not full. Concerning equivalence of institutions, it would be more pleasant not only to have two (simple) embeddings, but to have an embedding with Φ being an *equivalence of categories of theories*. However, this would rule out the above equivalences:

Proposition 4.8.2 Although they are simply equivalent in expressiveness, there is no simple embedding with Φ an equivalence of categories of theories between $P(\stackrel{f}{=}\Longrightarrow\stackrel{f}{=})$ and $R(R \Longrightarrow R =)$.

Proof: Let the mono-cardinality of an object in a category be the cardinality of the longest chain (if existing) of proper subobjects (i.e. monos which are not isos) ending in that object. If no proper subobject exists, the mono-cardinality is taken to be zero. Clearly, this notion is preserved by equivalences of categories. For example, in $\mathcal{SET}$, the mono-cardinality is the usual cardinality.

Then with the theory T_2 consisting of one sort, one unary relation symbol and no axioms, the category of sets with one relation can be specified with a $R(R \Longrightarrow R =)$-theory of mono-cardinality 2: In $R(R \Longrightarrow R =)$ T_2 has exactly two subtheories $T_0 \subseteq T_1 \subseteq T_2$, where T_0 is the empty theory and T_1 is the theory consisting of one sort symbol. But in $P(\stackrel{f}{=}\Longrightarrow\stackrel{f}{=})$, this works only with a theory having mono-cardinality 3: Let T_3 be the $P(\stackrel{f}{=}\Longrightarrow\stackrel{f}{=})$-theory

sorts s
ops $f\colon s \longrightarrow s$
$\forall x, y \colon s .\ f(x) \stackrel{f}{=} y \Longrightarrow f(x) \stackrel{f}{=} x$

Since T_3 consists of three components and each one is dependent on the previous components, it has exactly three subtheories in $P(\stackrel{f}{=}\Longrightarrow\stackrel{f}{=})$. Now the theories in $P(\stackrel{f}{=}\Longrightarrow\stackrel{f}{=})$ with mono-cardinality 2 consists of

- either two sort symbols
- or one sort and one operation symbol

- or one sort and one axiom

but all these theories do not specify the category of sets with one relation. □

Thus, requiring Φ to be an equivalence of categories of theories seems to be too strong.

The equivalence of proposition 4.8.1 is trivial, of course. The non-trivial task was to develop a suitable notion of equivalence.

Of course, the notions of embedding and of equivalence in expressiveness are rather strong. A weaker notion is developed in chapter 7.

Chapter 5

Five graphs of institutions

> "Categorists have developed a symbolism that allows one quickly to *visualize* quite complicated facts by means of *diagrams*."
> J. Adámek, H. Herrlich and G. Strecker [AHS90]

In the previous chapter, we have introduced three notions of institution representation [Mes89, Mos96a]:

1. Simple institution representations allow to map signatures to theories, sentences to sentences and models to models. Theorem-provers can be re-used along such representations (under some extra technical condition).

2. Conjunctive institution representations are like simple representations, except that a sentence may be translated to a finite set of sentences. Theorem-provers can still be re-used.

3. Weak institution representations map theories to theories and models to models. Under some condition, this can be made modular, such that sentences are mapped to theory extensions, and for mapping a theory, we then have to collect all the theory extensions. Semantic properties are kept along such representations, but since sentences are mapped to whole theory extensions, there is only a restricted re-use of theorem provers.

Each notion can be restricted to the case of embeddings, which means that signatures and sentences are embedded while model categories are linked via equivalences of categories. Equivalent expressiveness means embeddability in both directions. Now a systematic study of expressiveness of institutions should order them by their embeddability into each other.

In the following section, we establish five levels of expressiveness. At each level, several institutions are shown to be equivalent in expressiveness. Between institutions of different levels, there are only embeddings in one direction. Separation of the levels will be considered in chapter 6.

To give an impression of what these levels are, we here describe them informally, referring to some results that are shown later. The main distinction concerns the ability to conditionally generate equalities, relations, graphs of partial functions and data elements.

1. The first level is just total equational logic.

2. The second level adds relations, partiality and flat existence equations ($\stackrel{f}{=}$), while still staying equational.

3. The third level adds conditional equations, but only for total terms. Also equational partiality can be used at this level. At this level, one can conditionally generate equalities.

4. The fourth level adds predicates with Horn clauses, corresponding to conditional flat equations. This enables one to conditionally generate relations (and graphs of partial functions), but not data.

5. The fifth level introduces full conditional partial existence equations. This enables one to conditionally generate also data.

5.1 Equivalences among various institutions at level 5

In this section, we start with the most expressive level. We prove the equivalence of expressiveness of a number of institutions introduced in chapter 3 with respect to weak embeddings of institutions. This extends the following results: The equivalence of HEP-theories and left exact sketches is proved in [AR94] by showing that both have exactly the locally finitely presentable categories as model categories. Limit-theories are shown to have exactly locally finitely presentable categories as model categories in [Cos79]. In this section, we strengthen these results by constructing weak embeddings of institutions, which not only follow categorical constructions, but also act at the level of syntax. Furthermore, we give some new, perhaps even unexpected, results concerning other institutions. Partial algebras with strong equations are equivalent to partial algebras with universal Horn sentences. The exact relation of order-sorted algebra with sort constraints to partial algebras is clarified.

5.1. EQUIVALENCES AMONG VARIOUS INSTITUTIONS AT LEVEL 5

The main result gives us translations between all institutions under consideration. Thus it is a chance to unify the different branches of specification of partial algebras. Further, it is a step towards the goal of formal interoperability, multi-paradigm specification languages and re-use of theorem provers mentioned above.

Most of the results of this section have been announced in [Mos93] and published in [Mos96b].

Theorem 5.1.1 Between the institutions introduced in chapter 3, there are the institution representations shown in Fig. 5.1. Moreover, all of them except (5.14) preserve composable signatures, and all except (5.14), (5.17) and (5.19) have an isomorphism as model translation.

Proof:

The proof is given below, separately for each arrow in the diagram. Concerning preservation of composable signatures, note that the non-trivial representations translate signatures in a modular, component-wise manner, which makes the proof of left-adjointness used in Proposition 4.8.1 applicable. The trivial representations are inclusions of full subcategories on the signature level, which turn to out to be coreflective by taking as coreflection just forgetting those parts of the bigger signature that are not present in the smaller institution.[1] Since coreflective embeddings are left adjoints, preservation of colimits follows from Proposition A.2.3 in both cases. The only exception is (5.14): here, colimits are not preserved. The reason is that the subsort relation in $COS(=:\Rightarrow=:)$ is required to be transitive. This can lead to new injections in the representation of the colimit that are not present in the colimit of the representations. □

Corollary 5.1.2 All the institutions in Fig. 5.1 are weakly equivalently expressive when considered as rps pre-institutions. In particular, considered as specification frames, they are equivalently expressive.

Proof:

By composability of weak embeddings of rps pre-institutions. □

Corollary 5.1.3 All institutions in Fig. 5.1, except possibly $COS(=:\Rightarrow=:)$, have amalgamation.

Proof: By Theorem 3.9.4, $P(\stackrel{e}{=}\Rightarrow\stackrel{e}{=})$ has amalgamation. By the Theorem and Propositions 3.9.1, 3.9.2, 3.9.3 and 4.7.8, this carries over to the other institutions except $COS(=:\Rightarrow=:)$. □

---

[1]Note that this gives us an institution morphism in the sense of Chapter 8. Pairs of institution representations and institution morphisms arising by adjoint pairs of signature translations are studied in [AF96].

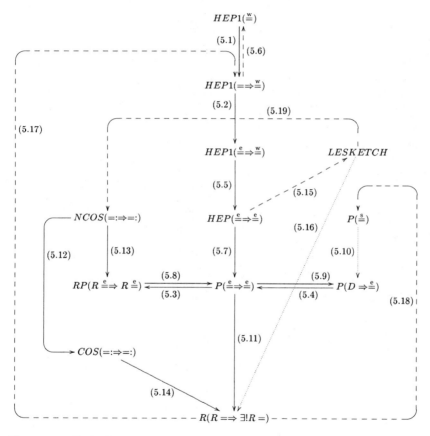

Figure 5.1: Embeddings among institutions of partial algebras. A solid arrow denotes a simple embedding of institutions, a dotted arrow denotes a conjunctive embedding of institutions, and a dashed arrow denotes a weak embedding of institutions.

5.1. Equivalences among various institutions at level 5

Corollary 5.1.4 All institutions in Fig. 5.1 are liberal.

Proof: $HEP(\stackrel{e}{=}\Rightarrow\stackrel{e}{=})$ has been shown to be liberal by Reichel [Rei87], and $RP(R \stackrel{e}{=}\Rightarrow R \stackrel{e}{=})$ is liberal by Fact 3.10.1. By the Theorem and Proposition 4.7.9, this carries over to the other institutions. □

Corollary 5.1.5 We can re-use theorem provers along all except the dashed arrows in Fig. 5.1.

Proof: It can be easily shown that all institutions under consideration have the property that model isomorphisms preserve satisfaction. Moreover, all model translations are equivalences of categories, so the are surjective up to isomorphism. So we can apply number 2 of Corollary 4.2.2. □

Corollary 5.1.6 (1) All model categories specifiable in any of the above institutions are locally finitely presentable.

(2) In all of the above institutions, the same subclass of locally finitely presentable categories can be specified.

(3) If we extend the above institutions to infinite signatures (consisting of infinite sets of sorts, infinite sets of (finitary) operation and relation symbols, and infinite graphs), then in each such extended institution, exactly the locally finitely presentable categories can be specified as model categories of theories.

Proof:

(1) See [Cos79, AR94, GU71] and use Corollary 5.1.2 and the fact that locally finite representability is invariant under equivalences of categories.

(2) Follows from Corollary 5.1.2.

(3) Coste [Cos79] shows the statement for $R(R \Longrightarrow \exists!R =)$, Adámek and Rosický [AR94] show it for HEP and $LESKETCH$. This carries over to the other institutions by noticing that the proof of Theorem 5.1.1 below can easily be extended to infinite signatures. □

Proof of Theorem 5.1.1, (5.1), (5.2), (5.3) and (5.4)

Obvious subinstitutions. □

Proof of Theorem 5.1.1, (5.5)

The simple institution representation $\mu\colon HEP1(\stackrel{\mathrm{e}}{=}\Rightarrow\stackrel{\mathrm{w}}{=}) \longrightarrow HEP1(\stackrel{\mathrm{e}}{=}\Rightarrow\stackrel{\mathrm{e}}{=})$ is defined by letting Φ and β_Σ be identities, while α_Σ maps a sentence

$$\forall X\,.\, t_1 \stackrel{\mathrm{e}}{=} u_1 \wedge \cdots \wedge t_n \stackrel{\mathrm{e}}{=} u_n \Longrightarrow t \stackrel{\mathrm{w}}{=} u$$

to

$$\forall X\,.\, t_1 \stackrel{\mathrm{e}}{=} u_1 \wedge \cdots \wedge t_n \stackrel{\mathrm{e}}{=} u_n \wedge t \stackrel{\mathrm{e}}{=} t \wedge u \stackrel{\mathrm{e}}{=} u \Longrightarrow t \stackrel{\mathrm{w}}{=} u$$

The representation condition easily follows from the definition of satisfaction for weak equations. □

Proof of Theorem 5.1.1, (5.6)

The weak embedding $\mu\colon HEP1(\!=\!\Rightarrow\stackrel{\mathrm{w}}{=}) \longrightarrow HEP1(\stackrel{\mathrm{w}}{=})$ is due to Rosický [AHR88]: Signatures, models and weak equations are left unchanged. A conditional Σ-equation $\varphi =$

$$\forall X\,.\, e_1 \wedge \cdots \wedge e_n \Longrightarrow t \stackrel{\mathrm{w}}{=} u$$

with $X = \{x_1 : s_1, x_n : s_n\}$ and t, u of sort s is translated to **derive from** T **by** $\Sigma \hookrightarrow T$, where T is the definitional extension of Σ by a partial operation $op^\varphi\colon s_1 \times \cdots \times s_n \longrightarrow s$ with domain condition

$$\mathit{Def}(op^\varphi) = X.\{\, e_i \mid i = 1,\ldots, n\,\}$$

and two weak equations

$$\forall X\,.\, op^\varphi(x_1,\ldots,x_n) \stackrel{\mathrm{w}}{=} t$$
$$\forall X\,.\, op^\varphi(x_1,\ldots,x_n) \stackrel{\mathrm{w}}{=} u$$

Now these two axioms together just state that, in case of definedness of $op^\varphi(x_1,\ldots,x_n)$ (that is, if the premises of φ hold) and of definedness of t and u, then $t = u$. Thus a Σ-model can be expanded to the extended theory iff it satisfies φ. □

5.1. Equivalences among various institutions at level 5

Proof of Theorem 5.1.1, (5.7)

We define a simple embedding of institutions $\mu^{hep}\colon HEP(\stackrel{e}{=}\!\Rightarrow\!\stackrel{e}{=}) \longrightarrow P(\stackrel{e}{=}\!\Rightarrow\!\stackrel{e}{=})$:

- $\Phi^{hep}(S, OP, POP, \preceq, \mathit{Def})$ is the theory that has the signature (S, OP, POP) and the following axioms (which are a circumscription of the $\mathit{Def}(pop)$-condition):

$$\forall X.\ pop(x_1, \ldots, x_n) \stackrel{e}{=} pop(x_1, \ldots, x_n) \Longrightarrow e \quad (X.e \in \mathit{Def}(pop))$$
$$\forall X.\ \bigwedge_{e \in \mathit{Def}(pop)} e \Longrightarrow pop(x_1, \ldots, x_n) \stackrel{e}{=} pop(x_1, \ldots, x_n)$$

for $pop\colon s_1, \ldots, s_n \longrightarrow s \in POP$ and $X = \{\, x_1 : s_1, \ldots, x_n : s_n \,\}$,

- $\Phi^{hep}(\sigma) = \sigma$,

- α_Σ^{hep} and β_Σ^{hep} identities. $\square$

Proof of Theorem 5.1.1, (5.8)

The embedding $\mu^{chardom}$ from section 4.8, representing relations by domains of partial operations, can easily be extended to a simple embedding of institutions $\mu^{chardom}\colon RP(R \stackrel{e}{=}\!\Rightarrow R \stackrel{e}{=}) \longrightarrow P(\stackrel{e}{=}\!\Rightarrow\!\stackrel{e}{=})$.

- $\Phi^{chardom}(S, OP, POP, REL)$ is the theory with signature

$$(S \cup \{\, one \,\},$$
$$OP \cup \{\, * : one \,\},$$
$$\{\, \chi^R \colon s_1, \ldots, s_n \longrightarrow one \mid R \colon s_1, \ldots, s_n \in REL \,\})$$

and the axiom $\forall x, y \colon one.\ x = y$.

- $\alpha_\Sigma^{chardom}$ does not change the variable system. On atomic formulas, the action is

$$\alpha_\Sigma^{chardom}\llbracket t_1 \stackrel{e}{=} t_2 \rrbracket = \llbracket t_1 \stackrel{e}{=} t_2 \rrbracket$$
$$\alpha_\Sigma^{chardom}\llbracket R(t_1, \ldots, t_n) \rrbracket = \llbracket \chi^R(t_1, \ldots, t_n) \stackrel{e}{=} * \rrbracket$$

This can be easily extended to conditional formulas.

- $\beta^{chardom}$ is as in section 4.8. $\square$

Proof of Theorem 5.1.1, (5.9)

The simple embedding of institutions $\mu^{pp}: P(\stackrel{e}{=}\Rightarrow\stackrel{e}{=}) \longrightarrow P(D \Rightarrow\stackrel{e}{=})$ is defined as follows: A signature $\Sigma = (S, OP, POP)$ is translated by Φ^{pp} to Σ extended by

ops $\chi^s_= : s \times s \longrightarrow ?s$
$\forall x:s . \; \chi^s_=(x,x) \stackrel{e}{=} x$
$\forall x,y:s . \; \chi^s_=(x,y) \stackrel{e}{=} \chi^s_=(x,y) \Longrightarrow x \stackrel{e}{=} y$

for each sort s in S.

An atomic formula $t_1 \stackrel{e}{=} t_2$ (of sort s) is translated by α^{pp} to

$$\chi^s_=(t_1, t_2) \stackrel{e}{=} \chi^s_=(t_1, t_2)$$

This translation is easily extended to conditional formulas.

β^{pp} translates a model by simply forgetting the $\chi^s_=$-operations. Since the latter are defined uniquely by the axioms, β^{pp} is an isomorphism of categories.

The representation condition follows from the fact that $\chi^s_=(x,y)$ is defined iff $x = y$. □

Proof of Theorem 5.1.1, (5.10)

The conjunctive embedding of institutions $\mu^{se}: P(\stackrel{s}{=}) \longrightarrow P(D \Rightarrow\stackrel{e}{=})$ is defined by:

- Φ^{se} and β^{se}_Σ are identities
- $\alpha^{se}_\Sigma(X : t_1 \stackrel{s}{=} t_2) = \{ \forall X . \; t_1 \stackrel{e}{=} t_1 \Longrightarrow t_1 \stackrel{e}{=} t_2, \forall X . \; t_2 \stackrel{e}{=} t_2 \Longrightarrow t_1 \stackrel{e}{=} t_2 \}$
  These axioms state that if either side of $t_1 \stackrel{s}{=} t_2$ is defined, the other side is defined as well and both sides are equal. Thus the representation condition follows. □

Proof of Theorem 5.1.1, (5.11)

The simple embedding of institutions $\mu^{limgra}: P(\stackrel{e}{=}\Rightarrow\stackrel{e}{=}) \longrightarrow R(R \Longrightarrow \exists! R =)$ is defined as follows:

- $\Phi^{limgra}(S, OP, POP)$ is the theory $((S, OP, REL), \Gamma)$, where REL con-

5.1. EQUIVALENCES AMONG VARIOUS INSTITUTIONS AT LEVEL 5

tains a relation symbol
$$G^{pop} : s_1, \ldots, s_n \, s$$
and Γ contains an axiom
$$\forall x_1 : s_1, \ldots, x_n : s_n; y, z : s \, . \\ G^{pop}(x_1, \ldots, x_n, y) \wedge G^{pop}(x_1, \ldots, x_n, z) \Longrightarrow y = z$$
for each $pop: s_1, \ldots, s_n \longrightarrow s \in POP$. The relation G^{pop} shall hold the graph of a partial operation pop, and the axiom states that G^{pop} is right-unique, that is, the graph of a partial operation.

- Following Burmeister [Bur82, p. 325], we translate existence equations in a relational form. $(t_1 \stackrel{e}{=} t_2)^*$ is defined inductively as follows:

  - $(pop(t_1, \ldots, t_n) \stackrel{e}{=} t_0)^* = (\exists y_0 : s_0 \ldots y_n : s_n \, . \, (G^{pop}(y_1, \ldots, y_n, y_0) \wedge (t_0 \stackrel{e}{=} y_0)^* \wedge \cdots \wedge (t_n \stackrel{e}{=} y_n)^*))$ for $pop: s_1 \ldots s_n \longrightarrow s_0 \in POP$
  - $(op(t_1, \ldots, t_n) \stackrel{e}{=} t_0)^* = (\exists y_0 : s_0 \ldots y_n : s_n \, . \, (op(y_1, \ldots, y_n) \stackrel{e}{=} y_0 \wedge (t_0 \stackrel{e}{=} y_0)^* \wedge \cdots \wedge (t_n \stackrel{e}{=} y_n)^*))$ for $op: s_1 \ldots s_n \longrightarrow s_0 \in OP$
  - $(x \stackrel{e}{=} t)^* = (t \stackrel{e}{=} x)^*$, if t is not a variable
  - $(x \stackrel{e}{=} y)^* = (x \stackrel{e}{=} y)$

  Considering a $P(\stackrel{e}{=} \Longrightarrow \stackrel{e}{=})$-axiom
  $$\varphi = \forall X \, . \, e_1 \wedge \cdots \wedge e_n \Longrightarrow e$$
  by the rules for prenex normal form from [Sho67], we can assume that e_i^* has form $\exists Y_i \, . \, \psi_i$ and e^* has form $\exists Y \, . \, \psi$, where the ψ_i and ψ consist of conjunctions of atomic formulas. Then
  $$\alpha_\Sigma^{limgra}(\varphi) = \forall X \cup Y_1 \cup \cdots \cup Y_n \, . \, \psi_1 \wedge \cdots \wedge \psi_n \Longrightarrow \exists ! Y \, . \, \psi$$

- β_Σ^{limgra} takes a $\Phi\Sigma$-model A and replaces each relation G_A^{pop} by the partial operation with graph G_A^{pop}.

Now the representation condition essentially is the proposition on page 326 of [Bur82], using the observation that
$$\forall X \, . \, (\exists Y_1 \, . \, \psi_1 \wedge \cdots \wedge \exists Y_n \, . \, \psi_n \Longrightarrow \exists Y \, . \, \psi)$$
is equivalent to
$$\forall X \, . \, \forall Y_1 \ldots \forall Y_n \, . \, (\psi_1 \wedge \cdots \wedge \psi_n \Longrightarrow \exists Y \, . \, \psi)$$
by the rules for prenex normal form from [Sho67], which in turn is equivalent to
$$\forall X \forall Y_1 \ldots \forall Y_n \, . \, (\psi_1 \wedge \cdots \wedge \psi_n \Longrightarrow \exists ! Y \, . \, \psi)$$
by the right-uniqueness axioms above. □

Proof of Theorem 5.1.1, (5.12)

Obvious subinstitution. □

Proof of Theorem 5.1.1, (5.13)

Define the simple embedding of institutions $\mu^{cr}\colon NCOS(=:\Rightarrow=:) \longrightarrow RP(R \stackrel{e}{=} \Rightarrow R \stackrel{e}{=})$ by

- $\Phi^{cr}(S, \leq, OP)$ is the theory

    sorts s for s the maximum of a connected component
    ops $op_{s_1,\ldots,s_n,s}\colon max(s_1) \times \cdots \times max(s_n) \longrightarrow ? \, max(s)$
    for $op\colon s_1, \ldots, s_n \longrightarrow s \in OP$,
    where $max(s)$ is the maximum of the connected component of s
    preds $R^s : max(s)$ for $s \in S$
    $\forall x\colon s \,.\, R^s(x)$ for s the maximum of a connected component
    $\forall x\colon max(s)\,.\, R^s(x) \Longrightarrow R^{s'}(x)$
    for $s \leq s'$ in $(S, \leq)$
    $\forall x_1\colon max(s_1), \ldots, x_n\colon max(s_n)\,.$
    $\quad op_{s_1,\ldots,s_n,s}(x_1, \ldots, x_n) \stackrel{e}{=} op_{s_1,\ldots,s_n,s}(x_1, \ldots, x_n)$
    $\quad \Longleftrightarrow \bigwedge_{i=1,\ldots,n} R^{s_i}(x_i)$
    for $op\colon s_1, \ldots, s_n \longrightarrow s \in OP$
    $\forall x_1\colon max(s_1), \ldots, x_n\colon max(s_n)\,.$
    $\quad \bigwedge_{i=1,\ldots,n} R^{s_i}(x_i)$
    $\quad \Longrightarrow op_{s_1,\ldots,s_n,s}(x_1, \ldots, x_n) \stackrel{e}{=} op_{s'_1,\ldots,s'_n,s'}(x_1, \ldots, x_n)$
    for $s_i \leq s'_i$ $(i = 1, \ldots, n)$ and $s \leq s'$

- Concerning sentence translation, conditinal equations are just translated by translating their constituent atomic formulas. An atomic formula

$$t = u$$

is translated to

$$LP_\Sigma(t) \stackrel{e}{=} LP_\Sigma(u)$$

and an atomic formula

$$t : s$$

is translated to

$$R^s(LP_\Sigma(t))$$

where $LP_\Sigma(t)$ is the *least sort parse* of t, defined in Proposition 3.8.1.

5.1. Equivalences among various institutions at level 5

- $\beta_{\Sigma}^{cr}(M')$ has as carrier for sort s the set $R_{M'}^{s}$ and as operations the $op_{M'}$ restricted to their domain.
  $\beta_{\Sigma}^{cr-1}(M)$ keeps just M_s as carrier M'_s (for s a maximum element of a connected component), and has no total operations. The op_M can be interpreted as partial operations on M', while the $R_{M'}^s$ (for $s \in S$) are simply the M_s.

Note that this construction just mimics the definition of an order sorted algebra, so the representation condition is easy. □

Proof of Theorem 5.1.1, (5.14)

Define the simple embedding of institutions $\mu^{ch}: COS(=:\Rightarrow=:) \longrightarrow R(R \Longrightarrow \exists!R =)$ by

- $\Phi^{ch}(\Sigma) = (\Sigma^{\#}, J)$ where for $\Sigma = (S, \leq, OP)$, $\Sigma^{\#}$ has sorts S, an operation symbol $op_{s_1,\ldots,s_n,s}: s_1, \ldots, s_n \longrightarrow s$ for $op: s_1, \ldots, s_n \longrightarrow s \in OP$ plus additional coercion operation symbols $c_{s,s'}: s \longrightarrow s'$ whenever $s \leq s'$. The axioms in J are

  (identity) $\forall x: s.\ c_{s,s}(x) = x$ for each $s \in S$
  (injectivity) $\forall x, y: s.\ c_{s,s'}(x) = c_{s,s'}(y) \Longrightarrow x = y$
  $\qquad$ for each $s \leq s' \in (S, \leq)$
  (transitivity) $\forall x: s.\ c_{s',s''}(c_{s,s'}(x)) = c_{s,s''}(x)$
  $\qquad$ for each $s \leq s' \leq s'' \in (S, \leq)$
  (homomorphism) $\forall x_1: s_1, \ldots, x_n: s_n$.
  $\qquad c_{s,s'}(op_{s_1,\ldots,s_n,s}(x_1, \ldots, x_n)) =$
  $\qquad op_{s'_1,\ldots,s'_n,s'}(c_{s_1,s'_1}(x_1), \ldots, c_{s_n,s'_n}(x_n))$
  $\qquad$ whenever $op: s_1, \ldots, s_n \longrightarrow s$ and $op: s'_1, \ldots, s'_n \longrightarrow s'$ are in OP
  $\qquad$ with $s_i \leq s'_i$ and $s \leq s'$

- As above, replace all sort constraints in the premises. Now

  – $\alpha_\Sigma(\forall X.\ t_1 = u_1 \wedge \cdots \wedge t_n = u_n \Longrightarrow t = u)$ is
  $$\forall X.\ LP_\Sigma(t_1) = LP_\Sigma(u_1) \wedge \cdots \wedge LP_\Sigma(t_n) = LP_\Sigma(u_n)$$
  $$\Longrightarrow LP_\Sigma(t) = LP_\Sigma(u)$$

  – $\alpha_\Sigma(\forall X.\ e_1, \wedge \cdots \wedge, e_n \Longrightarrow t: s)$ is
  $$\forall X.\ LP_\Sigma(t_1) = LP_\Sigma(u_1) \wedge \cdots \wedge LP_\Sigma(t_n) = LP_\Sigma(u_n) \Longrightarrow \exists! x: s.\ x = t$$
  where $x: s$ is a new variable not occurring in X.

- $\beta_\Sigma(B) = B^\bullet$, where coercions in B are replaced by set inclusions in $B^\bullet$ using colimits of filtered diagrams built up by the coercions. Goguen and Meseguer show β to be a pointwise equivalence of categories. The exact definition for $B^\bullet$ is given in [GM92].

The representation condition follows from Theorem 4.4 (2) of [GM92] and the coincidence of homomorphic extensions of valuations in B and $B^\bullet$. □

Proof of Theorem 5.1.1, (5.15)

The weak embedding of institutions $\mu^{hs}: HEP(\stackrel{e}{=}\Rightarrow\stackrel{e}{=}) \longrightarrow LESKETCH$ has already been sketched in [Poi86].

- $\Phi^{hs}(S, OP', POP', \preceq', Def)$ first resolves the overloading in (S, OP', POP') by renaming operation symbols in the same way as the construction of $\Sigma^{\#}$ in (5.14) does. Let $(S, OP, POP, \preceq, Def)$ be the result of this renaming. Then $\Phi^{hs}(S, OP', POP', \preceq', Def)$ is a sketch constructed as follows from $\Phi^{hs}(S, OP', POP', \preceq', Def)$:

    - A sort s leads to an object s in the sketch.
    - A total operation symbol $op: s_1, \ldots, s_n \longrightarrow s \in OP$ is translated to an arrow
    $$s_1 \times \cdots \times s_n \xrightarrow{op} s$$
    in the sketch, together with a limit cone for $s_1 \times \cdots \times s_n$ and the projections $\pi_i: s_1 \times \cdots \times s_n \longrightarrow s_i$ (in the sequel, we will assume that such limit objects, projections and cones are added without mentioning it).
    - By a well-founded induction along $\preceq$, we define the translation of a partial operation symbol pop, and of terms and existence equations containing only partial operation symbols less than or equal to pop. By induction hypothesis, within this definition, we can assume that the translation for partial operation symbols less than pop and of terms and existence equations containing partial operation symbols only less than pop has already been defined.
    Now given $pop \in POP$, terms $t \in T_\Sigma(X)$ ($X = \{x_1 : s_1; \ldots; x_n : s_n\}$), which consist of partial operation symbols less than pop and total operation symbols only, are translated to partial arrows, that is, pairs of arrows
    $$s_1 \times \cdots \times s_n \xleftarrow{\llbracket t \rrbracket^-} D_t \xrightarrow{\llbracket t \rrbracket^+} s$$

as described by Poigné [Poi86]. Since only partial operations less than pop are used in t, we can assume the existence of partial arrows corresponding to these operations when constructing $[\![t]\!]^+$ and $[\![t]\!]^-$. Now an existence equation $e = X.t_1 \stackrel{e}{=} t_2 \in \mathit{Def}(pop)$ is translated to $D_e \longrightarrow s_1 \times \cdots \times s_n$, which is the equalizer of $[\![t_1]\!]^+ \circ p$ and $[\![t_2]\!]^+ \circ q$, where p and q are the pullback of $[\![t_1]\!]^-$ and $[\![t_2]\!]^-$:

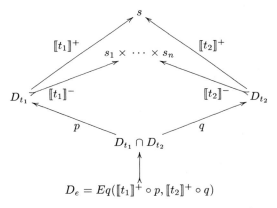

Now $pop\colon s_1, \ldots, s_n \longrightarrow s$ is translated to the partial arrow

$$s_1 \times \cdots \times s_n \xleftarrow{[\![pop]\!]^-} \mathrm{dom}\, pop = \bigcap_{e \in \mathit{Def}(pop)} D_e \xrightarrow{[\![pop]\!]^+} s$$

where the intersection is a categorical intersection, thus a limit.

- $\alpha_\Sigma^{hs}(\forall x_1\colon s_1; \ldots; x_n\colon s_n\,.\, e_1 \wedge \cdots \wedge e_k \Longrightarrow e)$ is **derive from** T **by** $\Sigma' \hookrightarrow T$, where $\Sigma' = sign(\Phi^{hs}(\Sigma))$ and T is $\Phi^{hs}(\Sigma)$ expanded by equalizers of pullbacks for $D_{e_1}, \ldots, D_{e_k}, D_e$, an arrow $m\colon D_{e_1} \cap \ldots \cap D_{e_k} \longrightarrow D_e$ and a diagram

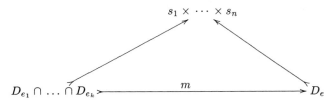

where uniqueness of m follows from the other two arrows being monic.

- β_Σ^{hs} maps a sketch model $M'\colon \Sigma' \longrightarrow |\,\mathbf{Set}\,|$ to the partial algebra with carriers $M'(s)$ ($s \in S$), total operations $M'(op)$ ($op \in OP$) and partial operations $M'(pop)\colon M'(\mathrm{dom}\, pop) \longrightarrow M'(s)$ ($pop\colon s_1, \ldots, s_n \longrightarrow s \in POP$).

A natural transformation $\eta\colon M' \longrightarrow M''$ is translated to the homomorphism of partial algebras $(\eta_s)_{s\in S}$.

Vice versa, given a partial Σ-algebra M, we construct a sketch model $\beta_\Sigma^{hs'} M$ by interpreting each object (resp. arrow) corresponding to a sort or domain (resp. operation symbol) by the corresponding set (resp. operation). The interpretation of the other parts of the sketch is then determined by the limit cones, if we chose a canonical limit structure on **Set**. Similarly for the homomorphisms. Then $\beta_\Sigma^{hs} \circ \beta_\Sigma^{hs'} = Id$, while $\beta_\Sigma^{hs'} \circ \beta_\Sigma^{hs}$ is an isomorphism.

The representation condition: Poigné's construction satisfies

$$M'(D_e \longrightarrow s_1 \times \cdots \times s_n) \cong \{\nu\colon X \longrightarrow \beta^{hs}(M') \mid \nu \models e\} \hookrightarrow \beta^{hs}(M'_{s_1}) \times \cdots \times \beta^{hs}(M'_{s_n})$$

Therefore, a model expansion of M' interpreting $m\colon D_{e_1} \cap \ldots \cap D_{e_k} \longrightarrow D_e$ exists if and only if

$$\bigcap_{i=1,\ldots,k} \{\nu\colon X \longrightarrow \beta^{hs}(M') \mid \nu \models e_i\} \subseteq \{\nu\colon X \longrightarrow \beta^{hs}(M') \mid \nu \models e\}$$

if and only if $\beta^{hs}(M') \models \forall x_1\colon s_1; \ldots ; x_n\colon s_n \,.\, e_1 \wedge \cdots \wedge e_k \Longrightarrow e$ □

Proof of Theorem 5.1.1, (5.16)

The conjunctive embedding of institutions $\mu^{sc}\colon LESKETCH \longrightarrow R(R \Longrightarrow \exists!R =)$ is defined by expressing the universal property of limits within $R(R \Longrightarrow \exists!R =)$.

$\Phi^{sc}(G,U)$ is the theory with a sort s for each object s in G and a total operation $f\colon s \longrightarrow s'$ for each arrow $f\colon s \longrightarrow s'$ in G. $\beta_{(G,U)}^{sc}(M')$ is the model $M\colon \Sigma \longrightarrow |\mathbf{Set}|$ with $M(s) = M'_s$ for objects s of G and $M(f) = f_{M'}$ for arrows f of G, the converse direction being defined analogously.

$\alpha_{(G,U)}^{sc}$ maps a diagram $I \xrightarrow{D} G$ to the set consisting of an axiom

$$\forall x\colon D_i \,.\, D_{m_n}(\ldots (D_{m_1}(x)) \ldots) = D_{r_k}(\ldots (D_{r_1}(x)) \ldots)$$

for any pair

5.1. Equivalences among various institutions at level 5

of paths in I.

Considering cones over a diagram $I \xrightarrow{D} G$, $(L, (\pi_i)_{i \in |I|})$ is a cone if in the interpretation, $D_m \circ \pi_i = \pi_j$ for each $m: i \longrightarrow j \in I$, and it is a limiting cone if for all other cones $(P, (p_i)_{i \in |I|})$ there exists a unique $p: P \longrightarrow L$ with $\pi_i \circ p = p_i$ for all $i \in |I|$.

This property is a second-order property: the existence of a unique **Set**-morphism, i. e. function p is required. But we can express this with first-order axioms by replacing cones $(P, (p_i)_{i \in |I|})$ by elements $(y_i = p_i(y) : D_i)_{i \in |I|}$ and maps $p: P \longrightarrow L$ by elements $x = p(y) : L$ (for a given $y : P$). Then the limiting property can be expressed with the following axioms:

$$\forall x: L \,.\, D_m(D_i(x)) = D_j(x) \text{ for } m: i \longrightarrow j \in I$$
$$\forall \{ y_i : D_i \mid i \in |I| \} \,.\, \bigwedge_{m: i \to j \in I} D_m(y_i) = y_j \Longrightarrow \exists! x: L \,.\, (\bigwedge_{i \in |I|} \pi_i(x) = x_i)$$

The set of these two axioms then is the translation by $\alpha^{sc}_{(G,U)}$ of a cone $(L, (\pi_i)_{i \in |I|})$ over $I \xrightarrow{D} G$. □

Proof of Theorem 5.1.1, (5.17)

We define a weak embedding $\mu^{rs} = (\Phi^{rs}, \alpha^{rs}, \beta^{rs}): R(R \Longrightarrow \exists! R =) \longrightarrow HEP1(\Longrightarrow \stackrel{w}{=})$. We have to represent relations from $R(R \Longrightarrow \exists! R =)$ within $HEP1(\Longrightarrow \stackrel{w}{=})$. The idea is that additional sorts R hold relations $R : s_1 \ldots s_n$. A sort R must be in some sense a "subsort" of $s_1 \times \cdots \times s_n$. Instead of introducing an injection from R to $s_1 \times \cdots \times s_n$ (which is possible only indirectly since we do not have product sorts), we introduce projections $\pi_i^R : R \longrightarrow s_i$ for $i = 1, \ldots, n$, which are "jointly injective", that is, $< \pi_1^R, \ldots, \pi_n^R > : R \longrightarrow s_1 \times \cdots \times s_n$ is injective. Then, R is a sort representing the relation of all tuples

$$(\pi_1^R(x), \ldots, \pi_n^R(x))$$

where x is a variable ranging over R.

Thus let Φ^{rs} extend a signature (S, OP, REL) by the following theory:

sorts R for each $R \in REL \cup \{ =_s \mid s \in S \}$
ops $\pi_i^R : R \longrightarrow s_i$ for $R : s_1 \ldots s_n \in REL \cup \{ =_s : s\, s \mid s \in S \}$, $i = 1, \ldots, n$
$\Delta_s : s \longrightarrow =_s$ for $s \in S$
$\forall x, y : R \,.\, \bigwedge_{i=1, \ldots, n} \pi_i^R(x) = \pi_i^R(y) \Longrightarrow x = y$
   for $R : s_1 \ldots s_n \in REL \cup \{ =_s : s\, s \mid s \in S \}$
$\forall x : s \,.\, \pi_i^{=_s}(\Delta_s(x)) = x$ for $s \in S$, $i = 1, 2$
$\forall y :=_s .\, \pi_1^{=_s}(y) = \pi_2^{=_s}(y)$ for $s \in S$

The first axiom scheme states that the π_i^R are "jointly injective", that is,

$< \pi_1^R, \ldots, \pi_n^R >$ is injective. For each $s \in S$, the other two axiom schemes (together with the diagonal function Δ_s) ensure that $=_s$ exactly is the equality relation on s.

To translate a $R(R \Longrightarrow \exists!R =)$-axiom φ, replace all equations $t_1 = t_2$ in φ by relation applications $=_s (t_1, t_2)$ to gain uniformity. Let the axiom thus modified be

$$\forall X \,.\, Q_1(t_1^1, \ldots t_{m_1}^1) \wedge \cdots \wedge Q_q(t_1^q, \ldots t_{m_q}^q) \implies$$
$$\exists!Y \,.\, (R_1(u_1^1, \ldots u_{n_1}^1) \wedge \cdots \wedge R_r(u_1^r, \ldots u_{n_r}^r))$$

(with $X = \{\, x_1 : s_1, \ldots, x_m : s_m \,\}$, $Y = \{\, y_1 : s_1', \ldots, y_n : s_n' \,\}$). In the sequel, assume the following conventions for variable ranges:

$$\begin{array}{ll} \iota = 1, \ldots, m & \kappa = 1, \ldots, n \\ i = 1, \ldots, q & k = 1, \ldots, r \\ j = 1, \ldots, m_i & l = 1, \ldots, n_k \end{array}$$

Now φ states that for all relation members of $Q_1, \ldots, Q_q$ of a certain form, there are relation members of $R_1, \ldots, R_r$ of a certain form depending on some additional variables Y. We want to simulate the effect of φ by introducing partial operations $proj\_R$, $proj\_y$ and $other\_y$. The first two of these have as arguments a valuation of the variables in X and relation members of $Q_1, \ldots, Q_q$. They have as values the corresponding relation members of $R_1, \ldots, R_r$ resp. the values of the variables in Y:

$$proj\_R_k^\varphi \colon s_1 \ldots s_m \; Q_1 \ldots Q_q \longrightarrow R_k \quad (k = 1, \ldots, r)$$

$$proj\_y_\kappa^\varphi \colon s_1 \ldots s_m \; Q_1 \ldots Q_q \longrightarrow s_\kappa' \quad (\kappa = 1, \ldots, n)$$

The ordering on these partial operations is discrete. Let $Q = \{\, v_1 : Q_1, \ldots, v_q : Q_q \,\}$ and $Def(proj\_R_k^\varphi) = Def(proj\_y_\kappa^\varphi) =$

$$X \cup Q . \{\, \pi_j^{Q_i}(v_i) = t_j^i \mid i = 1, \ldots, q, j = 1, \ldots, m_i \,\} \tag{5.1}$$

Thus the $proj\_R_k^\varphi$ and the $proj\_y_\kappa^\varphi$ are defined if and only if for $i = 1, \ldots, q$, the value of v_i represents a member of Q_i which makes the premise $Q_i(t_1^i, \ldots, t_{m_i}^i)$ true. This is the case if and only if for $x_1, \ldots, x_m$, the premises hold with relation members $v_1, \ldots, v_q$.

Next we have to state that if the premises hold, the conclusion holds as well. That is, $proj\_y_\kappa^\varphi(x_1, \ldots, x_m, v_1, \ldots, v_q)$ should yield some value which has the following property: When it is substituted for $y_\kappa : s_\kappa' \in Y$ ($\kappa = 1, \ldots, n$), $R_k(u_1^k, \ldots, u_{m_k}^k)$ becomes true, and $proj\_R_k^\varphi(x_1, \ldots, x_m, v_1, \ldots, v_q)$ represents the corresponding member of R_k ($k = 1, \ldots, r$). This is expressed by the following axioms:

$$\forall X \cup Q \,.\, \pi_l^{R_k}(proj\_R_k^\varphi(x_1, \ldots, x_m, v_1, \ldots, v_q)) \stackrel{w}{=} \omega_l^k \tag{5.2}$$

for $k = 1, \ldots, r$ and $l = 1, \ldots, n_k$, where ω_l^k is the term u_l^k with y_κ replaced by $proj\_y_\kappa^\varphi(x_1, \ldots, x_m, v_1, \ldots, v_q)$ for $\kappa = 1, \ldots, n$.

Now the $proj\_y_\kappa^\varphi(x_1, \ldots, x_m, v_1, \ldots, v_q)$ have to be the unique y_κ with the above property. This is expressed by the axioms

$$\forall X \cup Y \cup Q \cup R. \, \{\, \pi_l^{R_k}(w_k) = u_l^k \mid k = 1, \ldots, r, l = 1, \ldots, n_k \,\} \implies$$

$$proj\_y_\kappa^\varphi(x_1, \ldots, x_m, v_1, \ldots, v_q) \stackrel{w}{=} y_\kappa \tag{5.3}$$

for $\kappa = 1, \ldots, n$, where $R = \{\, w_1 : R_1, \ldots, w_r : R_r \,\}$.

Note that when passing from $proj\_y_\kappa^\varphi(x_1, \ldots, x_m, v_1, \ldots, v_q)$ to arbitrary y_κ, we have also to pass from $proj\_R_j^\varphi(x_1, \ldots, x_m, v_1, \ldots, v_q)$ to arbitrary w_k, since the w_k have the y_κ as components.

Now $\alpha_\Sigma^{rs}(\varphi)$ is **derive from** T **by** $\Sigma' \hookrightarrow T$, where $\Sigma' = sign(\Phi^{rs}(\Sigma))$ and T is the extension of $\Phi^{rs} \Sigma$ outlined above. We have to prove uniqueness of $\Sigma' \hookrightarrow T$-expansions (if existing) for Σ'-models M'. Let N and N' be two $\Sigma' \hookrightarrow T$-expansions of M'. Arguments for $proj\_y_\kappa^\varphi$ and $proj\_R_k^\varphi$ in N (resp. N') can be identified with valuations $\nu: X \cup Q \longrightarrow M'$. For such a valuation satisfying $Def(proj\_R_k^\varphi)$, extend ν to $\nu': X \cup Y \cup Q \cup R \longrightarrow N$ with $\nu'(y_\kappa) = proj\_y_{\kappa,N'}^\varphi(\nu)$ and $\nu'(w_k) = proj\_R_{k,N'}^\varphi(\nu)$. Since (5.2) holds in N', the premise of (5.3) holds for ν', so $proj\_y_{\kappa,N}^\varphi = proj\_y_{\kappa,N'}^\varphi$. By (5.2) and the injectivity of $<\pi_1^{R_k}, \ldots, \pi_{n_k}^{R_k}>$, we have also $proj\_R_{k,N}^\varphi = proj\_R_{k,N'}^\varphi$. Thus $N = N'$.

Considering models, $\beta_{(S,OP,REL)}^{rs}$ takes a Σ'-model M' and replaces for each relation symbol $R: s_1 \ldots s_n \in REL$ the components M_R' and $\pi_{i,M'}^R: M_R' \longrightarrow M_{s_i}'$ ($i = 1, \ldots, n$) by the relation $<\pi_{1,M'}^R, \ldots, \pi_{n,M'}^R> [M_R']$. Homomorphisms are left unchanged, except that the components for the R fall away. This can easily shown to be an equivalence of categories.

To prove the representation condition, assume that $M = \beta_\Sigma^{rs} M' \models_\Sigma \varphi$ with φ as above. To show that $M' \models_{\Sigma'}$ **derive from** T **by** $\Sigma' \hookrightarrow T$ (with $\Sigma' \hookrightarrow T$ as above), we have to construct a $\Sigma' \hookrightarrow T$-expansion N of M'. For simplicity, we assume that for $R: s_1, \ldots, s_n \in REL$, the injection $<\pi_{1,M'}^R, \ldots, \pi_{n,M'}^R>: M_R' \longrightarrow M_{s_1}' \times \cdots \times M_{s_n}'$ is an inclusion (without that, we would have to deal with some extra isomorphisms). For valuations $\nu: X \cup Q \longrightarrow M'$ satisfying $Def(proj\_R_k^\varphi)$, that is, $\pi_{j,M'}^{Q_i}(\nu(v_i)) = \nu^\#(t_j^i)$, we have to define $proj\_R_{k,N}^\varphi(\nu)$ and $proj\_y_{\kappa,N}^\varphi(\nu)$. Now by the equations satisfied by ν, $(\nu^\#(t_1^i), \ldots, (\nu^\#(t_{m_i}^i)) \in Q_{i,M'}$, that is, $\nu|_X$, the restriction of ν to X, satisfies the premises of φ. By the assumption, there is a unique extension $\nu': X \cup Y \longrightarrow M'$ of $\nu|_X$ satisfying the conclusion of φ. Now define

$$proj\_y_{\kappa,N}^\varphi(\nu) = \nu'(y_\kappa) \quad (\kappa = 1, \ldots, n)$$

$$proj\_R_{k,N}^\varphi(\nu) = ({\nu'}^\#(u_1^k), \ldots, {\nu'}^\#(u_{n_k}^k)) \quad (k = 1, \ldots, r)$$

This guarantees (5.2) and (5.3).

The other way round, assume that $M' \models_{\Sigma'}$ **derive from** T **by** $\Sigma' \hookrightarrow T$ holds. Assume further that a valuation $\nu\colon X \longrightarrow M = \beta_\Sigma^{rs} M'$ satisfies the premises of φ. Let N be the unique $\Sigma' \hookrightarrow T$-expansion of M'. Define $\xi\colon X \cup Q \longrightarrow N$ by $\xi(x_i) = \nu(x_i)$ and $\xi(v_i) = (\nu^{\#}(t_1^i), \ldots, \nu^{\#}(t_{m_i}^i))$. Then ξ satisfies $Def(proj\_R_k^\varphi)$, so $proj\_y_{\kappa,N}^\varphi$ and $proj\_R_{k,N}^\varphi$ are defined on ξ. Define $\nu'\colon X \cup Y \longrightarrow M$ by $\nu'(x_i) = \nu(x_i)$ and $\nu'(y_\kappa) = proj\_y_{\kappa,N}^\varphi(\xi)$. By (5.2), ν' satisfies the conclusion of φ. By (5.3), ν' is unique with that property. Thus φ holds in M. □

We illustrate how $\mu^{rs}\colon R(R \Longrightarrow \exists!R =) \longrightarrow \exists!(HEP1(\stackrel{e}{=}\Rightarrow\stackrel{w}{=}))$ works by giving an example. Consider the specifications of simple and transitive graphs in $R(R \Longrightarrow \exists!R =)$:

spec GRAPH1 =
  sorts *nodes*
  preds *Edge* : *nodes* × *nodes*

spec TRANSITIVE1 = **GRAPH1 then**
$\forall x, y, z\colon nodes\,.\; Edge(x,y) \wedge Edge(y,z) \Longrightarrow Edge(x,z)$

This is translated via μ^{rs} to $HEP1(\stackrel{e}{=}\Rightarrow\stackrel{w}{=})$ (up to a renaming and omission of $=_s$ and Δ_s which are needed only for equations):

spec GRAPH2 =
  sorts *nodes, edges*
  ops *source, target*: *edges* ⟶ *nodes*
    $\forall e, f\colon edges\,.\; source(e) = source(f) \wedge target(e) = target(f) \Longrightarrow e = f$

spec TRANSITIVE2 = **GRAPH2 then**
  ops $proj\_edge\colon nodes \times nodes \times nodes \times edges \times edges \longrightarrow ? edges$
    $Def(proj\_edge) = \{\, e, f : edges;\, x, y, z : nodes \,\}.$
      $\{\, source(e) = x, target(e) = y, source(f) = y, target(f) = z \,\}$
  $\forall e, f\colon edges;\, x, y, z\colon nodes\,.\; source(proj\_edge(x,y,z,e,f)) \stackrel{w}{=} x$
  $\forall e, f\colon edges;\, x, y, z\colon nodes\,.\; target(proj\_edge(x,y,z,e,f)) \stackrel{w}{=} z$

Since $\kappa = 0$ in this case, there are no axioms of form (5.3).

Now axiom (5.1) expresses the first three arguments of *proj_edge* in terms of the last two. Thus the redundant arguments of *proj_edge* can be omitted, and it becomes a concatenation operation:

spec TRANSITIVE3 = **GRAPH2 then**
  ops $\_\_\circ\_\_\colon edges \times edges \longrightarrow ? edges$
    $Def(\_\_\circ\_\_) = \{\, e, f : edges \,\}.\{\, source(f) = target(e) \,\}$

5.1. Equivalences among various institutions at level 5   91

$\forall e,f\!:\!edges\,.\ source(e\circ f) \stackrel{w}{=} source(e)$
$\forall e,f\!:\!edges\,.\ target(e\circ f) \stackrel{w}{=} target(f)$

By the way: using this representation of simple graphs, we can easily pass over to multigraphs by omitting the **GRAPH2**-axiom.

Of course, translations of complex theories may become very unreadable. They should better be viewed as the output of a compiler, which can be fed into (semi-)automatic tools.

Proof of Theorem 5.1.1, (5.18)

The construction of the weak embedding $\mu^{ls}\!:\!R(R \Longrightarrow \exists!R=) \longrightarrow P(\stackrel{s}{=})$ is similar to that of μ^{rs}. The new idea here is to represent relations R as domains of partial operations χ_R (with irrelevant values). Then an atomic relation application

$$R(t_1,\ldots,t_n)$$

is translated to

$$\chi_R(t_1,\ldots,t_n)\text{ is defined}$$

Conjunctions can be handled by nested terms with the outer operation symbol chosen to be total, since such a nested term is defined if and only if all subterms are defined. An equivalence

$$\forall X\,.\ R(t_1,\ldots,t_n) \Longleftrightarrow Q(u_1,\ldots,u_m)$$

can be translated to

$$\forall X\,.\ \chi_R(t_1,\ldots,t_n) \stackrel{s}{=} \chi_Q(u_1,\ldots,u_m)$$

Since implications $p \Longrightarrow q$ are equivalent to equivalences $p \Longleftrightarrow p \wedge q$ in propositional logic, we also can represent conditional axioms.

Now $\mu^{ls}\!:\!R(R \Longrightarrow \exists!R=) \longrightarrow P(\stackrel{s}{=})$ is defined as follows:

- $\Phi^{ls}(S,OP,REL)$ is (S,OP) extended by

    sorts $bool1$
    ops $true1:bool1$
        $first^s\!:\!s \times bool1 \longrightarrow? s$ for $s \in S$
        $\chi_R\!:\!s_1 \times \cdots \times s_n \longrightarrow? bool1$ for $R:s_1,\ldots,s_n \in REL$
        $\chi_{=_s}\!:\!s \times s \longrightarrow? bool1$ for $s \in S$
    $\forall x,y\!:\!bool1\,.\ x \stackrel{s}{=} y$
    $\forall x\!:\!s;y\!:\!bool1\,.\ first^s(x,y) \stackrel{s}{=} x$ for $s \in S$
    $\forall x\!:\!s\,.\ \chi_{=_s}(x,x) \stackrel{s}{=} true1$ for $s \in S$
    $\forall x,y\!:\!s\,.\ first^s(x,\chi_{=_s}(x,y)) \stackrel{s}{=} first^s(y,\chi_{=_s}(x,y))$

The first axiom ensures that $bool1$ is one-point (notice that $true1$ is a total operation), so the values of χ_R are irrelevant. The other axioms ensure that the domain of $\chi_{=_s}$ is the equality relation on s.

$\Phi^{ls}(\sigma)$ maps $bool1$ to $bool1$, $true1$ to $true1$, χ_R to $\chi_{\sigma(R)}$ and so on.

- Consider a $R(R \Longrightarrow \exists! R =)$-sentence $\varphi =$

$$\forall X . \, Q_1(t_1^1, \ldots t_{m_1}^1) \wedge \cdots \wedge Q_q(t_1^q, \ldots t_{m_q}^q) \Longrightarrow$$
$$\exists! Y . \, (R_1(u_1^1, \ldots u_{n_1}^1) \wedge \cdots \wedge R_r(u_1^r, \ldots u_{n_r}^r))$$

(with $X = \{x_1 : s_1, \ldots, x_m : s_m\}$, $Y = \{y_1 : s_1', \ldots, y_n : s_n'\}$ and equations $t_1 = t_2$ replaced by relation applications $=_s (t_1, t_2)$ to gain uniformity). To translate φ, introduce some auxiliary partial operations

ops $and_n \colon bool1^n \longrightarrow? \, bool1$
$and_q \colon bool1^q \longrightarrow? \, bool1$
$and_{r+1} \colon bool1^{r+1} \longrightarrow? \, bool1$
$forget^s \colon s \longrightarrow? \, bool1$ for $s \in S$
$\forall x_1, \ldots, x_n \colon bool1 . \, and_n(x_1, \ldots, x_n) \stackrel{s}{=} true1$
$\forall x_1, \ldots, x_q \colon bool1 . \, and_q(x_1, \ldots, x_q) \stackrel{s}{=} true1$
$\forall x_1, \ldots, x_{r+1} \colon bool1 . \, and_{r+1}(x_1, \ldots, x_{r+1}) \stackrel{s}{=} true1$
$\forall x \colon s . \, forget^s(x) \stackrel{s}{=} true1$ for $s \in S$

Further introduce partial operations

$$proj\_y_\kappa^\varphi \colon s_1 \ldots s_m \longrightarrow s_\kappa' \quad (\kappa = 1, \ldots, n)$$

As in the proof of (5.17), $proj\_y_\kappa^\varphi$ shall be defined if and only if its arguments, considered as valuation, make the premises of φ true. This is expressed by the axioms:

$$\forall X . \, and_q(\chi_{Q_1}(t_1^1, \ldots t_{m_1}^1), \ldots, \chi_{Q_q}(t_1^q, \ldots t_{m_q}^q)) \stackrel{s}{=}$$
$$forget_1^{s_\kappa'}(proj\_y_\kappa^\varphi(x_1, \ldots, x_m))$$

$(\kappa = 1, \ldots, n)$

Further, the conclusion shall hold in that case and the $proj\_y_\kappa^\varphi(x_1, \ldots, x_m)$ shall be unique with that property:

$$\forall X \cup Y . \, proj\_y_\kappa^\varphi(x_1, \ldots, x_m) \text{ defined} \Longrightarrow$$
$$(R_1(u_1^1, \ldots, u_{n_1}^1) \wedge \cdots \wedge R_r(u_1^r, \ldots, u_{n_r}^r))$$
$$\Longleftrightarrow y_1 = proj\_y_1^\varphi(x_1, \ldots, x_m) \wedge \cdots \wedge y_n = proj\_y_n^\varphi(x_1, \ldots, x_m))$$

Following the above discussion, this is translated to $P(\stackrel{s}{=})$-axioms

$\forall X \cup Y$.
$and_{r+1}(\chi_{R_1}(u_1^1, \ldots, u_{n_1}^1), \ldots, \chi_{R_r}(u_1^r, \ldots, u_{n_r}^r), proj\_y_1^\varphi(x_1, \ldots, x_m))$
$\stackrel{s}{=}$
$and_n(\chi_{=_{s_1'}}(y_1, proj\_y_1^\varphi(x_1, \ldots, x_m)), \ldots, \chi_{=_{s_n'}}(y_n, proj\_y_n^\varphi(x_1, \ldots, x_m)))$

Now $\alpha_\Sigma^{ls}(\varphi)$ is **derive from** T by $\Sigma' \hookrightarrow T$, where $\Sigma' = sign(\Phi^{ls}(\Sigma))$ and T is the extension of $\Phi^{ls}(\Sigma)$ outlined above.

- β_Σ^{ls} takes a Σ'-model M' and replaces the operations $\chi_{R,M'}$ by their domains (considered as relations), while the other operations introduced in T are forgotten. Homomorphisms are left unchanged.

  Vice versa, $\beta_\Sigma^{ls\,-1}$ replaces all relations R_M by partial operations $\chi_{R,M'}$ with the corresponding relation as domain. By the axiom $\forall x, y : bool1 . x \stackrel{s}{=} y$, the value of $\chi_{R,M'}$, if defined, is $true1_M$. The other operations introduced in T are determined by the axioms.

Now the representation condition can be verified as follows: Suppose $M = \beta_\Sigma^{ls} M' \models_\Sigma \varphi$ holds with φ as above. To show that $M' \models_{\Sigma'}$ **derive from** T by $\Sigma' \hookrightarrow T$ (with Σ' and T as above), we have to construct a $\Sigma' \hookrightarrow T$-expansion N of M'. Now $and_{n,N}$, $and_{q,N}$, $and_{r+1,N}$ and $forget_N$ are determined by the axioms. By noticing that $\chi_{R,N}(x_1, \ldots, x_n)$ is defined if and only if $R_{M'}(x_1, \ldots, x_n)$ holds, and $and_{n,N}$, $and_{q,N}$ and $and_{r+1,N}$ act as conjunction of definedness, we can proceed similarly to (5.17). □

Note that in $P(\stackrel{e}{=} \Rightarrow \stackrel{e}{=})$, we always can take $OP = \emptyset$ by regarding total operations as partial operations with $op(x_1, \ldots, x_n) \stackrel{e}{=} op(x_1, \ldots, x_n)$. μ^{limgra} translates this to a $R(R \Longrightarrow \exists ! R =)$-theory without total operations, and μ^{ls} further translates it into a $P(\stackrel{s}{=})$-theory with $OP = \{\, true1 : bool1 \,\}$. Thus, we need just one existence equation $true1 \stackrel{e}{=} true1$ to express $P(\stackrel{e}{=} \Rightarrow \stackrel{e}{=})$ or $R(R \Longrightarrow \exists ! R =)$ within $P(\stackrel{s}{=})$ *without* total operations. But this single existence equation is needed, since purely partial theories of strong equations all have the totally undefined algebra as a model, while this is not always the case in $P(\stackrel{e}{=} \Rightarrow \stackrel{e}{=})$.

Proof of Theorem 5.1.1, (5.19)

The weak institution representation $\mu^{sn} : LESKETCH \longrightarrow NCOS(=:\Rightarrow=:)$ is defined as follows:

- $\Phi^{sn}(G, U)$ is the theory with a sort s for each object s in G and a total operation $f\colon s \longrightarrow s'$ for each arrow $f\colon s \longrightarrow s'$ in G. The ordering on the sorts is discrete.

- $\alpha^{sn}_{(G,U)}$ maps a diagram $I \xrightarrow{D} G$ to **derive from** T **by** $\Sigma' \hookrightarrow T$, where $\Sigma' = sign(\Phi^{sn}(G,U))$ and T is $\Phi^{sn}(G,U)$ expanded by axioms

$$\forall x\colon D_i.\ D_{m_n}(\ldots(D_{m_1}(x))\ldots) = D_{r_k}(\ldots(D_{r_1}(x))\ldots)$$

for any pair

of paths in I.

- $\alpha^{sn}_{(G,U)}$ maps a limit cone $(L,(\pi_i)_{i\in|I|})$ over a diagram $I \xrightarrow{D} G$ to **derive from** T **by** $\Sigma' \hookrightarrow T$, where $\Sigma' = sign(\Phi^{sn}(G,U))$ and T is defined as follows: Let $\{i_1,\ldots,i_n\}$ be the objects of I. Define a sort $D_{i_1} \times \cdots \times D_{i_n}$ and operations $\hat{\pi}_{i_k}\colon D_{i_1} \times \cdots \times D_{i_n} \longrightarrow D_{i_k}$ $(k=1,\ldots,n)$ which shall hold the product of $D_{i_1},\ldots,D_{i_n}$ by the following theory expansion of $\Phi^{sn}(G,U)$:

sorts $D_{i_1} \times \cdots \times D_{i_n}$
ops $<\_,\ldots,\_>\colon D_{i_1} \ldots D_{i_n} \longrightarrow D_{i_1} \times \cdots \times D_{i_n}$
$\hat{\pi}_{i_k}\colon D_{i_1} \times \cdots \times D_{i_n} \longrightarrow D_{i_k})$ $(k=1,\ldots,n)$
$\forall x\colon D_{i_1} \times \cdots \times D_{i_n}.\ <\hat{\pi}_{i_1}(x),\ldots,\hat{\pi}_{i_n}(x)> = x$
$\forall x_1\colon D_{i_1};\ldots;x_n\colon D_{i_n}.\ \hat{\pi}_{i_k}(<x_1,\ldots,x_n>) = x_k$ $(k=1,\ldots,n)$

Now limits are equalizers of products, and equalizers are subsets in **Set**. This is expressed by requiring

$$L \leq D_{i_1} \times \cdots \times D_{i_n}$$

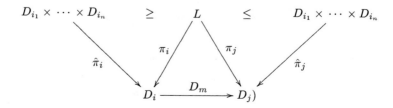

Following the construction of limits by equalizers of products described in theorem 12.3 of [AHS90], the projections of the limit shall be the projections of the product (composed with the above subsort inclusion):

$$\forall y\colon L \,.\, \pi_i(y) = \hat{\pi}(y)$$

and the subset L shall extract exactly those points of the product where the arrows of the limit cone commute with the projections. One inclusion of this set equation is expressed by equations

$$\forall y\colon L \,.\, D_m(\hat{\pi}_i(y)) = \hat{\pi}_j(y) \text{ for } m\colon i \longrightarrow j \in I$$

and the other one by the conditional sort constraint

$$\forall x\colon D_{i_1} \times \cdots \times D_{i_n} \,.\, \bigwedge_{m\colon i \longrightarrow j \in I} D_m(\hat{\pi}_i(x)) = \hat{\pi}_j(x) \implies x : L$$

- $\beta^{sn}_{((G,U),\emptyset)}(M')$ is the model $M\colon \Sigma \longrightarrow |\mathbf{Set}|$ with $M(s) = M'_s$ for objects s of G and $M(f) = f_{M'}$ for arrows f of G, the converse direction being defined analogously.

The representation condition can be verified with theorem 12.3 of [AHS90]. □

5.2 Categorical intuitionistic type theory

Intuitionistic type theory with a topos-theoretical semantics is developed by Lambek and Scott [LS86]. It should be mentioned here, without giving a formal definition, because it is a liberal institution and therefore is somewhat related to the other institutions studied above. Surprisingly, it can be embedded with a simple embedding into $P(\stackrel{e}{=\!=}\!\Rightarrow\!\stackrel{e}{=\!=})$!

I briefly sketch the construction of the embedding. By [BW85], the category of topoi and strict logical functors can be specified within $LESKETCH$, and therefore, by Theorem 5.1.1, also in $P(\stackrel{e}{=\!=}\!\Rightarrow\!\stackrel{e}{=\!=})$. Now translate a signature of categorical intuitionistic type theory to the specification of the category of topoi extended by constants for the types and the arrows in the signature and equations specifying source and target of arrows. From a model of this theory, a topos and an interpretation of the language over the signature in the topos can be extracted, this is the model's translation to a model of categorical intuitionistic type theory. Now sentence translation is easy be using the definition of satisfaction of sentences on page 143 of [LS86].

Moreover, at least some of the liberal categorical logics in the sense of Meseguer [Mes89] seem to be embeddable in $P(\stackrel{e}{=\!=}\!\Rightarrow\!\stackrel{e}{=\!=})$ as well.

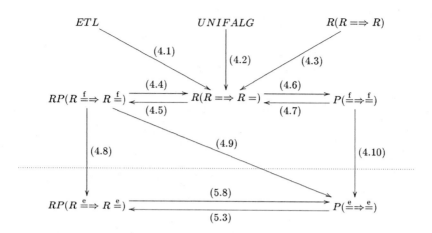

Figure 5.2: Embeddings at level 4, and selected relations to level 5. A solid arrow denotes a simple embedding of institutions.

5.3 Equivalences at level 4, and embeddings to level 5

Theorem 5.3.1 Between the institutions introduced in chapter 3, there are the institution representations shown in Fig. 5.2.

Proof:
The simple equivalence of $R(R \Longrightarrow R =)$ and $P(\stackrel{f}{=}\Rightarrow \stackrel{f}{=})$ has been shown already in section 4.8. So we are done with (4.6) and (4.7) in Figure 5.2. (4.4) is an easy extension of (4.7), which just leaves relations and relation applications unchanged.

(4.1), (4.2), (4.3), (4.5) (4.8) and (4.10) are obvious subinstitutions. (4.9) is the composition of (4.4), (4.6) and (4.10). □

5.4 Equivalences at level 3, and embeddings to level 4

Theorem 5.4.1 Between the institutions introduced in chapter 3, there are the institution representations shown in Fig. 5.3.

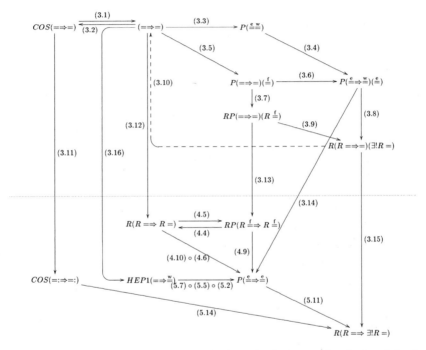

Figure 5.3: Embeddings at level 3, and selected relations to levels 4 and 5. A solid arrow denotes a simple embedding of institutions, and a dashed arrow denotes a weak embedding of institutions.

Proof of Theorem 5.4.1, (3.1)

The proof proceeds similar to the proof of (5.14). Define the simple embedding of institutions $\mu^{ch}: COS(=\Rightarrow=) \longrightarrow (=\Rightarrow=)$ by

- $\Phi^{ch}(\Sigma) = (\Sigma^\#, J)$

- $\alpha_\Sigma(\forall X . \; t_1 = u_1 \wedge \cdots \wedge t_n = u_n \Longrightarrow t = u)$ is

$$\forall X . \; LP_\Sigma(t_1) = LP_\Sigma(u_1) \wedge \cdots \wedge LP_\Sigma(t_n) = LP_\Sigma(u_n) \\ \Longrightarrow LP_\Sigma(t) = LP_\Sigma(u)$$

- $\beta_\Sigma(B) = B^\bullet$

The representation condition follows from Theorem 4.4 (2) of [GM92] and the coincidence of homomorphic extensions of valuations in B and $B^\bullet$. □

Proof of Theorem 5.4.1, (3.2)

In order to define a simple embedding $\mu: (=\Rightarrow=) \longrightarrow COS(=\Rightarrow=)$, we cannot take the "obvious" subinstitution because in $(=\Rightarrow=)$, we have unrestricted overloading (in particular, overloading of constants), while in $COS(=\Rightarrow=)$, there is only a restricted form of overloading (in particular, no overloading of constants). Instead, we have to decorate operation symbols with their arities and coarities.

Formally, translate a signature $\Sigma = (S, OP)$ to $\Phi(\Sigma) = (S, =, \{\, op_{w,s}: w \longrightarrow s \mid op: w \longrightarrow s \in OP \,\})$.

A translation of Σ-terms to pairs, consisting of a $\Phi(\Sigma)$-term and a $\Phi(\Sigma)$-sort, is defined inductively by translating a variable $x \in X_s$ to the pair (x, s) and translating $op(t_1, \ldots, t_n) : s$ by first inductively translating t_i to (u_i, s_i) and then taking $(op_{s_1 \ldots s_n, s}(u_1, \ldots, u_n), s)$ as the translation of $op(t_1, \ldots, t_n) : s$. This translation can now be easily extended to sentences by forgetting the sort component.

Models are not affected by the translation.

The representation condition follows from the observation that we only have renamed operation symbols. □

Proof of Theorem 5.4.1, (3.3)

The simple embedding of ($\Longrightarrow$=) into $P(\stackrel{\mathrm{e}}{=}\stackrel{\mathrm{w}}{=})$ exploits the fact that satisfaction of weak equations is defined with a conditional statement (if both sides are defined, they are equal) to encode general conditional equations. Therefore, equations have to be encoded into definedness conditions with the technique used in the proof of (5.18).

$\Phi(S, OP)$ is the signature (S, OP) extended by

ops $first: s \times s' \longrightarrow s$ for $s, s' \in S$
$\quad G^{op}: s_1 \times \cdots \times s_n \times s \longrightarrow ? \, s$ for $op: s_1, \ldots, s_n \longrightarrow s \in OP$
$\forall x: s, y: s' \, . \, first(x, y) = x$ for $s, s' \in S$
$\forall x_1: s_1, \ldots x_n: s_n, y: s \, . \, G^{op}(x_1, \ldots, x_n, op(x_1, \ldots, x_n)) \stackrel{\mathrm{e}}{=} op(x_1, \ldots, x_n)$
$\forall x_1: s_1, \ldots x_n: s_n, y: s \, . \, G^{op}(x_1, \ldots, x_n, y) \stackrel{\mathrm{w}}{=} y$
$\forall x_1: s_1, \ldots x_n: s_n, y: s \, . \, G^{op}(x_1, \ldots, x_n, y) \stackrel{\mathrm{w}}{=} op(x_1, \ldots, x_n)$

By the second axiom, the domain of G^{op} contains the graph of the operation op. By the third and the fourth axiom, the domain of G^{op} is contained in the graph of op. Furthermore, by the third axiom, the value of G^{op} is determined to be the projection to the value component of the graph.

α_Σ translates a Σ-sentence

$$\forall X \, . \, op^1(t_1^1, \ldots, t_{m_1}^1) = u_1 \wedge \cdots \wedge op^q(t_1^q, \ldots, t_{m_q}^q) = u_q \Longrightarrow t = u$$

to the $\Phi(\Sigma)$-sentence

$$\forall X \, . \, t \stackrel{\mathrm{w}}{=}$$
$$first(u, first(G^{op^1}(t_1^1, \ldots, t_{m_1}^1, u_1), first(\ldots, G^{op^q}(t_1^q, \ldots, t_{m_q}^q, u_q) \ldots)))$$

β_Σ takes a $\Phi(\Sigma)$-model A' and simply forgets the operations $G^{op}_{A'}$ and $first_{A'}$, while β_Σ^{-1} extends a Σ-model A with new operations G^{op}_A and $first_A$ which are uniquely determined by the axioms.

The representation condition can be proved as follows:
$\forall X \, .$
$t \stackrel{\mathrm{w}}{=} first(u, first(G^{op^1}(t_1^1, \ldots, t_{m_1}^1, u_1), first(\ldots, G^{op^q}(t_1^q, \ldots, t_{m_q}^q, u_q) \ldots)))$
holds in a $\Phi(\Sigma)$-model A' iff for all valuations $\nu: X \longrightarrow A'$, definedness of $G^{op^1}(t_1^1, \ldots, t_{m_1}^1, u_1), \ldots, G^{op^q}(t_1^q, \ldots, t_{m_q}^q, u_q)$ under ν implies $\nu \models t = u$ iff for all valuations $\nu: X \longrightarrow A'$, $\nu \models op^1(t_1^1, \ldots, t_{m_1}^1) = u_1 \wedge \cdots \wedge op^q(t_1^q, \ldots, t_{m_q}^q) = u_q$ implies $\nu \models t = u$ iff for all valuations $\nu: X \longrightarrow \beta_\Sigma(A')$, $\nu \models op^1(t_1^1, \ldots, t_{m_1}^1) = u_1 \wedge \cdots \wedge op^q(t_1^q, \ldots, t_{m_q}^q) = u_q$ implies $\nu \models t = u$ iff $\beta_\Sigma(A')$ satisfies $\forall X \, . \, op^1(t_1^1, \ldots, t_{m_1}^1) = u_1 \wedge \cdots \wedge op^q(t_1^q, \ldots, t_{m_q}^q) = u_q \Longrightarrow t = u$. $\square$

Proof of Theorem 5.4.1, (3.4), (3.5), (3.6), (3.7)

Obvious subinstitutions. □

Proof of Theorem 5.4.1, (3.8)

The simple embedding of institutions $\mu\colon P(\stackrel{e}{=}\!\!\Rightarrow\!\!\stackrel{w}{=})(\stackrel{e}{=}) \longrightarrow R(R \Longrightarrow\!\!=)(\exists!R =)$ is a combination of a restriction and a modification of the embedding (5.19).

Note that $\Phi^{(5.19)}(\Sigma)$ only contains conditional axioms with equational conclusion. A sentence

$$\forall X \,.\, t_1 \stackrel{e}{=} t_2$$

is mapped by (5.19) to an unconditional unique-existential statement. To translate a sentence

$$\forall X \,.\, t_1 \stackrel{e}{=} u_1 \wedge \cdots \wedge t_n \stackrel{e}{=} u_n \Longrightarrow t_0 \stackrel{w}{=} u_0$$

let $\exists Y_i \,.\, \varphi_i$ be $(y_i \stackrel{e}{=} t_i)^*$ and $\exists Z_i \,.\, \psi_i$ be $(z_i \stackrel{e}{=} u_i)^*$ $(i = 0, \ldots, n)$. Then the translation is

$$\forall X \cup \bigcup_{i=0,\ldots,n} (Y_i \cup Z_i) \,.\, \bigwedge_{i=0,\ldots,n} (\varphi_i \wedge \psi_i) \wedge \bigwedge_{i=1,\ldots,n} y_i = z_i \Longrightarrow y_0 = z_0$$

The proof of the representation condition is similar to that for (5.19). □

Proof of Theorem 5.4.1, (3.9)

$(3.8) \circ (3.6)$ is a simple embedding of $P(\Longrightarrow\!\!=)(\stackrel{f}{=})$ into $R(R \Longrightarrow\!\!=)(\exists!R =)$, and $R(R)$ is an obvious subinstitution of $R(R \Longrightarrow\!\!=)(\exists!R =)$. By combining these two, we get a simple embedding of $RP(\Longrightarrow\!\!=)(R \stackrel{f}{=})$ into $R(R \Longrightarrow\!\!=)(\exists!R =)$. □

Proof of Theorem 5.4.1, (3.10)

The weak embedding $\mu\colon R(R \Longrightarrow\!\!=)(\exists!R =) \longrightarrow (\Longrightarrow\!\!=)$ is a restriction of the weak embedding $\mu^{rs}\colon R(R \Longrightarrow \exists!R =) \longrightarrow HEP1(\stackrel{e}{=}\!\!\Rightarrow\!\!\stackrel{w}{=})$ numbered (5.17).

Note that signatures are translated by Φ^{rs} to theories which are already in ($\Longrightarrow$=). Axioms of form $\forall X \exists! Y \,.\, (e_1 \wedge \cdots \wedge e_n)$ are translated by μ^{rs} to sentences in **derive!**($=\!\!\Rightarrow\!\!=$), since, in the notation of the proof of (5.17), $q = 0$, thus the operations $proj\_R_k^\varphi$, $proj\_y_\kappa^\varphi$ are total. Axioms of form

$$\forall X \,.\, Q_1(t_1^1, \ldots, t_{m_1}^1) \wedge \cdots \wedge Q_q(t_1^q, \ldots, t_{m_q}^q) \Longrightarrow u_1 = u_2$$

are translated to

$$\forall X \cup V \,.\, \bigwedge_{i=1,\ldots,q,\, j=1,\ldots,m_i} \pi_j^{Q_i}(v_i) = t_j^i \Longrightarrow u_1 = u_2$$

This axiom has as premise the definedness condition of $proj\_R_k^\varphi$, while the conclusion directly states the desired condition. Thus the proof of the representation condition for (5.17) can be easily adapted. □

Proof of Theorem 5.4.1, (3.11), (3.12), (3.13)

Obvious subinstitutions. □

Proof of Theorem 5.4.1, (3.14)

The simple institution representation $\mu\colon P(\stackrel{e}{=}\!\!\Rightarrow\!\!\stackrel{w}{=})(\stackrel{e}{=}) \longrightarrow P(\stackrel{e}{=}\!\!\Rightarrow\!\!\stackrel{e}{=})$ is defined similarly to (5.5) by letting Φ and β_Σ be identities, and α_Σ maps a sentence

$$\forall X \,.\, t_1 \stackrel{e}{=} u_1 \wedge \cdots \wedge t_n \stackrel{e}{=} u_n \Longrightarrow t \stackrel{w}{=} u$$

to

$$\forall X \,.\, t_1 \stackrel{e}{=} u_1 \wedge \cdots \wedge t_n \stackrel{e}{=} u_n \wedge t \stackrel{e}{=} t \wedge u \stackrel{e}{=} u \Longrightarrow t \stackrel{w}{=} u$$

while an existence equation $\forall X \,.\, t_1 \stackrel{e}{=} t_2$ is left unchanged. The representation condition follows from the definition of weak satisfaction. □

Proof of Theorem 5.4.1, (3.15), (3.16)

Obvious subinstitutions. □

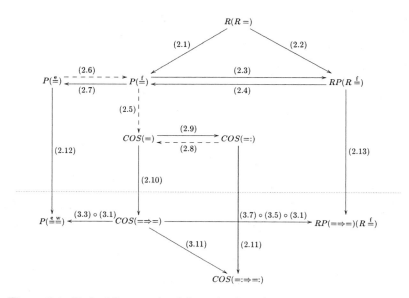

Figure 5.4: Embeddings at level 2, and selected relations to level 3. A solid arrow denotes a simple embedding of institutions, and a dashed arrow denotes a weak embedding of institutions.

5.5 Equivalences at level 2, and embeddings to level 3

Theorem 5.5.1 Between the institutions introduced in chapter 3, there are the institution representations shown in Fig. 5.4.

Proof of Theorem 5.5.1, (2.1)

Restrict (4.6) to the unconditional case. □

Proof of Theorem 5.5.1, (2.2), (2.3)

Obvious subinstitutions. □

Proof of Theorem 5.5.1, (2.4)

Restrict (4.6) ∘ (4.4) to the unconditional case. □

Proof of Theorem 5.5.1, (2.5)

Define the weak embedding $\mu: P(\stackrel{f}{=}) \longrightarrow COS(=)$ by

- $\Phi(\Sigma)$ with $\Sigma = (S, OP, POP)$ is $(S, =, OP)$ plus, for each $pop: s_1, \ldots, s_n \longrightarrow s \in POP$ the following theory:

  sorts $dom^{pop} \leq s_1 \times \cdots \times s_n$
  ops $<\_\_, \ldots, \_\_>: s_1 \times \cdots \times s_n \longrightarrow s_1 \times \cdots \times s_n$
  $\pi_i: s_1 \times \cdots \times s_n \longrightarrow s_i \ (i = 1, \ldots, n)$
  $val^{pop}: dom^{pop} \longrightarrow s$
  $\forall x_1: s_1, \ldots, x_n: s_n \,.\, \pi_i(<x_1, \ldots, x_n>) = x_i$
  $\forall x: s_1 \times \cdots \times s_n \,.\, <\pi_1(x), \ldots, \pi_n(x)> = x$

- for $\varphi = \forall x_1: s_1, \ldots, x_n: s_n \,.\, pop(t_1, \ldots, t_n) \stackrel{f}{=} t$, $\alpha_\Sigma(\varphi)$ is **derive from** T **by** $\Sigma \hookrightarrow T$, where T is the definitional extension of $\Phi(\Sigma)$ by

  ops $op^\varphi: s_1 \times \cdots \times s_n \longrightarrow dom^{pop}$
  $\forall x_1: s_1, \ldots, x_n: s_n \,.\, op^\varphi(x_1, \ldots, x_n) = <t_1, \ldots, t_n>$
  $\forall x_1: s_1, \ldots, x_n: s_n \,.\, val^{pop}(op^\varphi(x_1, \ldots, x_n)) = t$

  Since the newly introduced operation is defined explicitly, we have made an extension by definitions, which guarantees uniqueness of expansions.

- $\beta_\Sigma(M')$ just replaces each new sort dom^{pop} together with operation val^{pop} by a partial operation with domain dom^{pop} and taking as values those of val^{pop} (while forgetting the product structure). This replacement can easily be reversed, so we get an isomorphism of categories.

To verify the representation condition, note that it is always possible to expand a $\Phi(\Sigma)$-model M' by defining op^φ as required by the first axiom. This expansion then satisfies the second axiom iff $\beta_\Sigma(M') \models_{sign(\Phi(\Sigma))} \varphi = \forall x_1: s_1, \ldots, x_n: s_n \,.\, pop(t_1, \ldots, t_n) \stackrel{f}{=} t$, because we may substitute $<t_1, \ldots, t_n>$ for $op^\varphi(x_1, \ldots, x_n)$. □

Proof of Theorem 5.5.1, (2.6)

Define the weak embedding $\mu: P(\stackrel{e}{=}) \longrightarrow P(\stackrel{f}{=})$ to be the identity on signatures and models. Let $\Sigma = (S, OP, POP)$. A Σ-sentence

$$\forall X \, . \, t_1 \stackrel{e}{=} t_2$$

with $X = \{ x_1 : s_1, \ldots x_n : s_n \}$ is translated to **derive from** T **by** $\Sigma \hookrightarrow T$, where T is the extension of Σ by the following:

ops $f^t: s_1, \ldots, s_n \longrightarrow s$ for t a subterm of t_1 or t_2 having sort s
$\forall X \, . \, f^{x_i}(x_1, \ldots, x_n) = x_i$ for $i = 1, \ldots, n$
$\forall X \, . \, f^{op(u_1, \ldots, u_m)}(x_1, \ldots, x_n) = op(f^{u_1}(x_1, \ldots, x_n), \ldots, f^{u_m}(x_1, \ldots, x_n))$
     for $op: s_1, \ldots, s_m \longrightarrow s \in OP$, $op(u_1, \ldots, u_m)$ a subterm of t_1 or t_2
$\forall X \, . \, f^{pop(u_1, \ldots, u_m)}(x_1, \ldots, x_n) \stackrel{f}{=} pop(f^{u_1}(x_1, \ldots, x_n), \ldots, f^{u_m}(x_1, \ldots, x_n))$
     for $pop: s_1, \ldots, s_m \longrightarrow s \in POP$, $pop(u_1, \ldots, u_m)$ a subterm of t_1 or t_2
$\forall X \, . \, f^{t_1}(x_1, \ldots, x_n) = f^{t_2}(x_1, \ldots, x_n)$

Since this is an extension by definition, model expansions are unique.

To prove the representation condition, suppose that $A \models \forall X \, . \, t_1 \stackrel{e}{=} t_2$. For a valuation $\nu: X \longrightarrow A$, define $f^t_{A'}$ on ν to be $\nu^{\#}(t)$, which is defined since t is a subterm of t_1 or t_2. Then first three axioms above hold by definition of $\nu^{\#}$, and the forth axiom holds since $\nu^{\#}(t_1) = \nu^{\#}(t_2)$. Thus we have an expansion A' of A to the above theory.

Conversely, suppose that there is an expansion A' of A to the above theory, and let $\nu: X \longrightarrow A$ be a valuation. By the first three axioms, $f^t_{A'}$ on ν is $\nu^{\#}(t)$. By the forth axiom, $\nu^{\#}(t_1) = \nu^{\#}(t_2)$. □

Proof of Theorem 5.5.1, (2.8)

Define the weak embedding $\mu: COS(=:) \longrightarrow COS(=)$ by

- Φ and β are identities.

- α_Σ leaves a Σ-sentence $\forall X \, . \, t_1 = t_2$ unchanged and maps a Σ-sentence $\varphi = \forall X \, . \, t : s$ (with $X = \{ x_1 : s_1, \ldots x_n : s_n \}$) to **derive from** T **by** $\Sigma \hookrightarrow T$, where T is the definitional extension of Σ by

    ops $op^\varphi: s_1, \ldots, s_n \longrightarrow s$
    $\forall X \, . \, op^\varphi(x_1, \ldots, x_n) = t$

5.6. Level 1, and embeddings to level 2

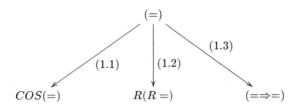

Figure 5.5: Embeddings at level 1. A solid arrow denotes a simple embedding of institutions.

Now a Σ-model A has an expansion to the extension iff op_A on a valuation $\nu\colon X \longrightarrow A$ can be taken to be $\nu^{\#}(t)$ iff for all valuations $\nu\colon X \longrightarrow A$, $\nu^{\#}(t)$ belongs to A_s iff A satisfies $\forall X \,.\, t : s$. □

Proof of Theorem 5.5.1, (2.7), (2.9), (2.10), (2.11), (2.12), (2.13)

Obvious subinstitutions. □

5.6 Level 1, and embeddings to level 2

Theorem 5.6.1 Between the institutions introduced in chapter 3, there are the institution representations shown in Fig. 5.5.

Proof:
Obvious subinstitutions. □

5.7 Rewriting logic

Meseguer's *rewriting logic* [Mes92] at the first glance seems to fit well in the collection of institutions introduced in chapter 3. However, concerning the question of embeddability by institution representations, rewriting logic behaves very strangely. In fact, using the notion of embedding developed in Chapter 4, only equational logic is embeddable into rewriting logic, since there are no conditional equations in rewriting logic and conditional rewrites do have a different semantics. Thus, it seems that rewriting logic is a universal framework for representing *entailment systems*, but not *institutions* (at least if a neat representation of model categories is desired).

Furthermore, rewriting logic is not embeddable into any of the institutions studied in this chapter, because it is not liberal. To see this, consider the rewriting theory consisting of one sort s, two constants $a, b : s$ and one rewrite $a \Rightarrow b$. Models of this theory are categories $\mathbf{A}$ with two distinguished objects A and B such that there exists a morphism from A to B. Model morphisms are functors $F\colon \mathbf{A} \longrightarrow \mathbf{A}'$ preserving the distinguished objects A and B, *but not necessarily the morphism from A to B*.

Assume that $(\mathbf{A}, A, B)$ is initial in this category of models. Let $\mathbf{C}$ be the category consisting of two objects A and B (which serve as distinguished objects), and (apart from the identities) two morphisms $f, g\colon A \longrightarrow B$. By initiality, there is a functor $F\colon \mathbf{A} \longrightarrow \mathbf{C}$ preserving A and B. Now define $F'\colon \mathbf{A} \longrightarrow \mathbf{C}$ to be F on objects, but on morphisms, it computes the value of F for the morphism and then swaps f and g. Since there is a morphism from A to B in $\mathbf{A}$, $F \neq F'$, contradicting initiality.

Thus rewriting logic should be studied together with general first order theories, which is beyond the scope of this thesis.

However, there is a specification frame, also called rewriting logic by Meseguer, which is different from $\mathbf{Th_0}$ applied to the institution called rewriting logic, and which *is* liberal. It has the same theories and models as the institution rewriting logic, but allows only those model morphisms that preserve the natural transformations required to exist by the rewrites in the theory. Clearly, this does not come from an institution, since model morphisms are restricted in presence of sentences, while in the institutional framework, theories are considered to specify *full* subcategories of those models satisfying the sentences.

The specification frame rewriting logic indeed is embeddable easily into, say, $P(\stackrel{e}{=}\Rightarrow\stackrel{e}{=})$ by noting that the theory of small categories can be specified in $P(\stackrel{e}{=}\Rightarrow\stackrel{e}{=})$ (see [Rei87]).

Chapter 6

Hierarchy theorems

> There was an old main who said 'Do
> Tell me how I'm to add two and two?
> Arithmetical lore
> Claims they add up to four
> But I fear that is almost too few.'[1]

The previous chapter was devoted to the study of embeddings and equivalences among institutions. Thus the realm of institutions introduced in chapter 3 is now ordered according to expressive power. Note that, up to weak equivalence, we already have even a total ordering on the institutions (if one does not count special restrictions like ETL).

In this chapter, some negative results will be proved, yielding a strict hierarchy of institutions and showing that the different levels introduced in the previous chapter really are different levels of expressiveness.

Hierarchy theorems are usually proved by finding some property that holds for all objects in some class, but that does not hold for some objects in a more expressive class, which then turns out to be strictly more expressive.

Here, we use

1. properties of the model categories of institutions, to show that some model category definable in a more expressive institution is not definable in a less expressive institution, and

2. properties of specifiable parameterized abstract data types (PADTs), where specifiable PADTs are considered to be functors specifiable using

---
[1]From J. Dahl: 99 Limericks. Langewiesche-Brandt 1962.

import, body and export specifications in the sense of [EM90]. Technically, these are free functors followed by forgetful functors.

Both approaches lead to a hierarchy of five levels. Part of the PADT hierarchy is also set up in [Mosa] using recursion theoretical methods. Here, we get simplified proofs by directly using the usual construction of free PADTs.

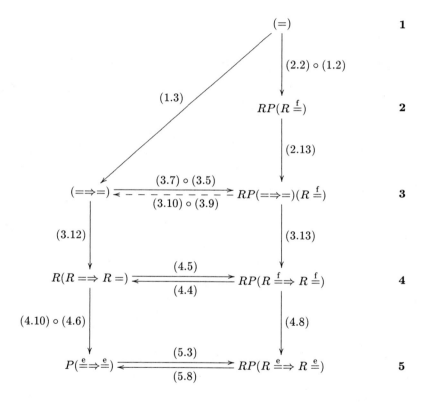

The usefulness of the hierarchy theorem is illustrated by locating a sample PADT (bounded stacks) in the hierarchy. In general, given a sample PADT, the algebraic properties help to find out the lowest position (= most restricted institution) in the hierarchy where it can be specified. This is important because the available tools may become weaker, if we choose a too general institution.

The main part of this chapter was published in [Mos95], and some part in [Mosa].

From the five graphs and levels introduced chapter 5, we choose one or two institutions per level to get the following overview (the institutions on the left hand side are well-known in the literature, while the equivalent institutions on the right hand-side form a more uniform line since four of them have relations and partiality).

We now examine the hierarchy $(=)$, $RP(R \stackrel{f}{=})$, $RP(=\Rightarrow=)(R \stackrel{f}{=})$, $RP(R \stackrel{f}{=} \Rightarrow R \stackrel{f}{=})$, $RP(R \stackrel{e}{=} \Rightarrow R \stackrel{e}{=})$ (that is, one representative per equivalence class w. r. t. weak equivalence). We prove properties of the institutions that can be used to separate the levels of the hierarchy from each other. We start with properties of model categories, then look at initial semantics and finally at parameterized abstract data types.

6.1 A model theoretic hierarchy theorem

Theorem 6.1.1 The five institutions we consider form a proper hierarchy:

	Institution	Properties of model categories	Separating example (constructed below)
1	$(=)$	has effective equivalence relations	
2	$RP(\stackrel{f}{=} R)$	subobjects commute with coequalizers	graphs = binary relations
3	$RP(=\Rightarrow=)(R \stackrel{f}{=})$	forgetful functors preserve regular epis	some special theory
4	$RP(R \stackrel{f}{=} \Rightarrow R \stackrel{f}{=})$	(regular epi,mono)-factorizable	transitive binary relations
5	$RP(R \stackrel{e}{=} \Rightarrow R \stackrel{e}{=})$	locally finitely presentable	transitive multigraphs or small categories

The proof is split into four parts, each of which is presented in one of the subsequent subsections.

6.1.1 Level 5 versus level 4: Partial Conditional Logic and Horn Clause Logic

The essential difference between level 4 and level 5 is that in the former, the theorem of homomorphisms holds, while in the latter, it fails. Categorically, the theorem of homomorphisms means that there exist (regular epi, mono)-factorizations.

Proposition 6.1.2 $RP(R \stackrel{e}{=} \Rightarrow R \stackrel{e}{=})$ and $R(R \Rightarrow R =)$ are not equivalent in expressiveness, that is, $RP(R \stackrel{e}{=}\Rightarrow R \stackrel{e}{=})$ is strictly more expressive than $R(= R \Rightarrow = R)$.

Proof:

For a theory $T = (\Sigma, \Gamma)$ in $RP(R \stackrel{e}{=}\Rightarrow R \stackrel{e}{=})$ and a model morphism $f: A \longrightarrow B$ in $\mathbf{Mod}(T)$, we can try to get a factorization of f through its image by the following procedure: first, take the kernel of f, that is, the pullback

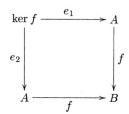

(By Corollary 5.1.6 and Proposition A.3.3, $\mathbf{Mod}(T)$ is complete and cocomplete.) Then, take the coequalizer of e_1 and e_2 to get the image $f[A]$:

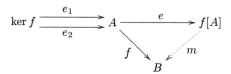

By the universal property of the coequalizer, there exists a unique $m: f[A] \longrightarrow B$ with $f = m \circ e$.

By remarks 3.4(2) and 5.13(2) of [AR94], if T belongs to $R(= R \Rightarrow = R)$, e is surjective, which implies that m is injective, hence a monomorphism, and (e, m) is a (regular epi,mono)-factorization of f. This is the well-known theorem of homomorphisms, generalized to $R(= R \Rightarrow = R)$.

In contrast to this, in $RP(R \stackrel{e}{=}\Rightarrow R \stackrel{e}{=})$, e may be not surjective and m no monomorphism, that is, there is no (regular epi,mono)-factorization of f. In [Rei87, 3.4], a counterexample very similar to Example 6.4.1 is discussed in detail. For factorizations in $P(\stackrel{e}{=}\Rightarrow\stackrel{e}{=})$ and $RP(R \stackrel{e}{=}\Rightarrow R \stackrel{e}{=})$, we have to iterate the construction of kernels and coequalizers (taking e instead of f and so on) possibly infinitely often.

Now (regular epi,mono)-factorizability clearly is preserved by equivalences of categories. So there cannot be an institution representation from $RP(R \stackrel{e}{=}\Rightarrow R \stackrel{e}{=})$ to $R(= R \Rightarrow = R)$ with model translation being a pointwise equivalence of categories. □

Corollary 6.1.3 $RP(R \stackrel{e}{=}\Rightarrow R \stackrel{e}{=})$ and $P(\stackrel{e}{=}\Rightarrow\stackrel{e}{=})$ are strictly more expressive than $RP(R \stackrel{f}{=}\Rightarrow R \stackrel{f}{=})$ and $R(R =\Rightarrow R =)$. □

6.1.2 Level 4 versus level 3: Horn Clause Logic and Conditional Equational Logic

The difference between level 3 and level 4 is that in the latter, there are quotient homomorphisms that are not closed, while in the former all homomorphisms are closed. Categorically this means, that forgetful functors preserve regular epis.

Proposition 6.1.4 $R(R =\Rightarrow R =)$ is not a subinstitution of $(=\Rightarrow =)$. That is, $R(R =\Rightarrow R =)$ is strictly more expressive than $(=\Rightarrow =)$.

Proof: A difference between $R(R =\Rightarrow R =)$ and $(=\Rightarrow =)$ is that in $R(R =\Rightarrow R =)$, there are quotients which are not closed. This cannot happen in $(=\Rightarrow =)$, where all homomorphisms are closed. Consider the theories

spec BINREL =
  sorts s
  preds $\leq: s \times s$

spec TRANSITIVE = **BINREL then**
$\forall x, y, z : s . x \leq y \wedge y \leq z \Longrightarrow x \leq z$

Now in **TRANSITIVE**, if we identify the elements b and c in

$$A = \{ a \leq b;\ c \leq d \}$$

we get the quotient

$$Q = \{ [a] \leq [b,c] \leq [d] \}$$

with $[a] \leq [d]$. The quotient homomorphism $q : A \longrightarrow Q$ with $q(x) = [x]$ is not closed: we have $q(a) \leq q(d)$, but not $a \leq d$. On the other hand, in **BINREL** the quotient $q' : A \longrightarrow Q'$ with $Q' = \{ [a] \leq [b,c] \leq [d] \}$ *is* closed, since we do not have $q'(a) \leq q'(d)$ in Q'.

To be able to use this for the theory of institution representations, we have to switch to a categorical formulation. We observe that q is a regular epimorphism in **TRANSITIVE**, but not in **BINREL** (since it does not factor through q'). So the forgetful functor **Mod** σ associated to the inclusion $\sigma:$ **BINREL** $\longrightarrow$ **TRANSITIVE** does not preserve regular epis.

On the other hand, in $(=\Rightarrow =)$, the regular epis are precisely the surjective homomorphisms (this follows from [AHS90], 7.72(1), 23.39 and 24.9). Since all forgetful functors in $(=\!\!\Rightarrow\!\!=)$ preserve surjectivity, they also preserve regular epis.

Now, preservation of regular epis by forgetful functors is a categorical property inherited to subinstitutions. So $R(R =\!\!\Rightarrow R =)$ cannot be a subinstitution of $(=\!\!\Rightarrow\!\!=)$.

Other categorical properties used to separate $(=\!\!\Rightarrow\!\!=)$ from $R(R =\!\!\Rightarrow R =)$ use essentially the same intuition. In $(=\!\!\Rightarrow\!\!=)$, all model categories have a dense set of regular projectives, whereas $R(R =\!\!\Rightarrow R =)$ lacks this property (see [AR94, 3.19, 3.21]; see [Bar89] for a similar property).

Corollary 6.1.5 $P(\stackrel{f}{=}\!\!\Rightarrow\!\stackrel{f}{=})$ and $RP(R \stackrel{f}{=}\!\!\Rightarrow R \stackrel{f}{=})$ cannot be subinstitutions of $RP(=\!\!\Rightarrow\!\!=)$ $(R \stackrel{f}{=})$ or $(=\!\!\Rightarrow\!\!=)$ either. □

6.1.3 Level 3 versus level 2: Conditional Equational Logic and Partial Equational Logic

The difference between level 3 and level 2 is the presence of conditional axioms at level 3. This implies that the construction of quotients (categorically: coequalizers) is more complex: We cannot just take the algebraic quotient, but have to factor it by new equations which may be derived using the conditional axioms from the elements of the congruence relation. In contrast to this, at level 2, quotients are just algebraic quotients.

To be able to formulate this categorically, we need to compare two different quotient constructions, one of which is done on a subalgebra. Now at level 2, construction of quotients commutes with subalgebras in the following sense: Given a congruence relation R on an algebra A and a subalgebra A' of A, then R induces a congruence relation R' on A'. Then A'/R' is a subalgebra of A/R in a natural way, while at level 3, this is not the case.

Definition 6.1.6 For a category having coequalizers and pullbacks, we say that *subobjects commute with coequalizers* in this category, if the following holds: Given an equivalence relation $R \underset{q}{\overset{p}{\rightrightarrows}} A$ on an object A and a subobject

6.1. A MODEL THEORETIC HIERARCHY THEOREM

$m\colon A' \longrightarrow A$ of A (i. e. m is a monomorphism), taking the pullback

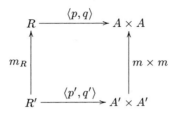

and then taking coequalizers $R \underset{q}{\overset{p}{\rightrightarrows}} A \overset{e}{\longrightarrow} Q$ and $R' \underset{q'}{\overset{p'}{\rightrightarrows}} A' \overset{e'}{\longrightarrow} Q'$ has the result that the unique factorization $m_Q\colon Q \longrightarrow Q'$ of $e \circ m$ through e' (which exists by the universal property of the coequalizer) *is a monomorphism*.

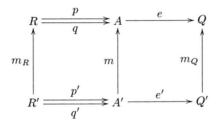

Proposition 6.1.7 For all theories $T = (\Sigma, \Gamma)$ in $P(\overset{e}{=})$, in **Mod**(T), subobjects commute with coequalizers.

Proof:

Let $R \underset{q}{\overset{p}{\rightrightarrows}} A$ be a relation on A and $m\colon A \longrightarrow A'$ be a subobject of A. By Proposition 3.6.2, coequalizers are just quotients. Thus m_Q is given by

$$m_Q([a]_{\langle p',q' \rangle}) = [m(a)]_{\langle p,q \rangle}$$

for $a \in A'$. By Proposition 3.6.3, $R' \underset{q'}{\overset{p'}{\rightrightarrows}} A'$ is $R \underset{q}{\overset{p}{\rightrightarrows}} A$ restricted to A'.
Thus, m_Q is injective. □

A counterexample showing that in model categories of $(\Longrightarrow=)$, generally subobjects do not commute with coequalizers, is the following. Let $T = (\Sigma, \Gamma)$ be the theory

```
spec T =
    sorts   s
    ops     f : s → s
            a, b : → s
    ∀x : s . f(x) = x ⇒ a = b
```

Let A be the T-algebra

```
algebra A =
    Carriers
        |A|_s = { 1, 2, 3, 4 }
    Functions
        a_A = 1
        b_A = 2
        f_A(1) = 2
        f_A(2) = 1
        f_A(3) = 4
        f_A(4) = 3
```

and A' be the subalgebra on the set $\{1, 2\}$. Now let $\equiv$ be the congruence generated by $3 \equiv 4$. Then $A/\equiv$ does not satisfy the T-axioms; in order to do so, we further have to identify 1 and 2 by taking $F_{(\Sigma,\emptyset) \rightarrowtail (\Sigma,\Gamma)}(A/\equiv)$. This then is the corresponding coequalizer, having carrier set $\{[1,2],[3,4]\}$. $\equiv$ restricted to A is the diagonal relation $\Delta A'$, and $A'/\Delta A' = A'$ is the corresponding coequalizer. The unique homomorphism $m_Q \colon A'/\Delta A' \longrightarrow F_{(\Sigma,\emptyset) \rightarrowtail (\Sigma,\Gamma)}(A/\equiv)$ is not injective, since it identifies 1 and 2. □

Corollary 6.1.8 ($=\Rightarrow=$) and $RP(=\Rightarrow=)(\overset{f}{=})$ are neither subinstitutions of $P(\overset{f}{=})$ nor of $P(\overset{e}{=})$.

6.1.4 Is there a subhierarchy within level 2?

While at all other levels, all institutions on one level are weakly equivalent in expressiveness (if one does not count special restrictions such as ETL), this is not the case for level 2. Indeed, at level 2, there may be up to three sublevels of increasing expressiveness w. r. t. to weak embeddings:

- Level 2a, consisting of $R(R=)$.
- Level 2b, consisting of $P(\overset{f}{=})$, $RP(R\overset{f}{=})$ and $P(\overset{e}{=})$.
- Level 2c, consisting of $COS(=)$ and $COS(=:)$.

I conjecture that these three sublevels can be separated by using finer categorical properties related to equivalence relations.

6.1.5 Level 2 versus level 1: Partial Equational Logic and Total Equational Logic

The difference between level 2 and level 1 concerns the role of congruence relations: while at level 1, the algebraic structure of a congruence relation (considered as an algebra) is determined already by the set-theoretical relation, at level 2, there is more flexibility, i. e. there are several algebras with the same congruence relation as carrier, but only one of them is "effectively" generated as a kernel of some homomorphism.

Proposition 6.1.9 In $(=)$, each model category has effective equivalence relations (for the category-theoretic definition of effective equivalence relation, see Definition A.4.1).

Proof:
See Remarks 3.4(8) and 3.6(7) in [AR94]. □

Example 6.1.10 Consider the signature Graph consisting of one sort symbol and one binary relation symbol. This is clearly a signature in $RP(\stackrel{f}{=} R)$, even in $R(R =)$. In the sequel, we identify the models of Graph with graphs. Now **Mod(Graph)**, the category of graphs, does not have effective equivalence relations: Consider the graph G

Then $G \times G$ is

and $\Delta G \subseteq G \times G$, the diagonal $\langle id_G, id_G \rangle \colon G \longrightarrow G \times G$ is

Now consider the relation R on G:

It is given by the inclusion $m: R \longrightarrow G \times G$, and is represented by the pair $\langle \pi_1 \circ m, \pi_2 \circ m \rangle$. Now ΔG is contained in R, so $R \xrightarrow{m} G$ is reflexive.

Define $flip: R \longrightarrow R$ by just flipping $(1,0)$ and $(0,1)$. Then $flip$ is an isomorphism from the subobject of G represented by $\langle \pi_1 \circ m, \pi_2 \circ m \rangle$ to the subobject represented by $\langle \pi_2 \circ m, \pi_1 \circ m \rangle$. Thus $R \xrightarrow{m} G$ is symmetric.

Composing $R \xrightarrow{m} G$ with itself by pulling back yields the diagonal relation (as a subobject of $G \times G \times G$, but this plays no role), which is contained in $R \xrightarrow{m} G$ by reflexivity above. Thus $R \xrightarrow{m} G$ is transitive.

Now suppose that for some graph homomorphism $f: G \longrightarrow G'$ the inner square in

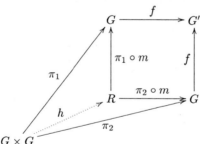

is a pullback. Since m is the identity on the underlying sets, the outer diagram commutes as well. By the pullback property, there has to exists a graph homomorphism $h: G \times G \longrightarrow R$ with $\pi_1 \circ m \circ h = \pi_1$ and $\pi_2 \circ m \circ h = \pi_2$. By the uniqueness property of the product, $m \circ h = id$, so h has like m to be the identity on the underlying sets. But this violates the graph homomorphism property.

Therefore, $R \xrightarrow{m} G$ is an equivalence relation which cannot be given by a kernel pair. Thus **Mod(Graph)** does not have effective equivalence relations. □

Corollary 6.1.11 Neither $RP(R \xrightarrow{f})$ nor $R(R =)$ is a subinstitution of $(=)$.

Proof:

Clearly, effective equivalence relations are preserved by equivalences of categories. □

A very similar counterexample can be constructed in the category of unary relations, but I found the above example more instructive. In general, relations introduce some kind of flexibility: information which is not represented by data elements. This kind of flexibility is present in $RP(\stackrel{f}{=} R)$ and $R(R =)$, but not in $(=)$.

6.2 Initial semantics

The results of the previous section concern the problem of specifying model categories (loose semantics). Another common problem is the question whether some given abstract data type (ADT) is specifiable with initial semantics. Here, semi-computable ADTs and specifiable ADTs (with hidden machinery) turn out to be the same. An ADT is semi-computable iff it is (up to isomorphism) a quotient (w.r.t. a recursively enumerable congruence) of an algebra consisting of recursive sets of numbers as carriers and (partial) recursive operations. An ADT is specifiable ADTs with hidden machinery in an institution, if it is a reduct of an initial algebra of a theory in that institution. For more detailed definitions, see [BT87].

Theorem 6.2.1 Concerning initial semantics, the hierarchy collapses almost totally: In $RP(R \stackrel{f}{=})$, $RP(=\Rightarrow=)(R \stackrel{f}{=})$, $RP(R \stackrel{f}{=} \Rightarrow R \stackrel{f}{=})$ and $RP(R \stackrel{e}{=} \Rightarrow R \stackrel{e}{=})$ (resp. $(=)$), the ADTs specifiable initially using hidden machinery coincide with the semi-computable ADTs (resp. total semi-computable ADTs).

Institution	Initially specifiable ADTs (with hidden machinery)
$(=)$	Total semi-computable ADTs
$RP(R \stackrel{f}{=})$	Semi-computable ADTs
$RP(=\Rightarrow=)(R \stackrel{f}{=})$	Semi-computable ADTs
$RP(R \stackrel{f}{=} \Rightarrow R \stackrel{f}{=})$	Semi-computable ADTs
$RP(R \stackrel{e}{=} \Rightarrow R \stackrel{e}{=})$	Semi-computable ADTs

Proof: See [Mosa]. □

6.3 Properties of parameterized ADTs

Parameterized abstract data types (PADTs) are useful for designing modular specifications. Specification with hidden machinery here means specification of modules with import, export and body parts in the sense of Ehrig and Mahr [EM90]. The body ADT is constructed freely over the parameter ADT, and the export interface is a view that forgets all irrelevant details ("hidden parts") of the body ADT.

Consider now the question whether some PADT can be specified with hidden machinery, i.e. whether it is expressible as composite of a free and a forgetful functor (see section 2.7):

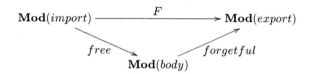

The following theorem was proved in [Mosa] with recursion theoretic methods. We now prove it more directly, using the construction of parameterized abstract data types (PADTs) in fact 3.10.1. Before doing that, we need to introduce a technical notion, namely what it means that a PADT functor F preserves kernels. Preservation of kernels roughly means that F of a kernel should yield a kernel again. But since a PADT functor F applied to a kernel does not even have the form of a binary relation, we have to be careful with the exact formulation.

Definition 6.3.1 Let a PADT functor F be given. Let A^2 be the product of A with itself, with projections $\pi_1, \pi_2: A^2 \longrightarrow A$. Let $\Pi: F A^2 \longrightarrow (F A)^2 := \ < F \pi_1, F \pi_2 >$. Let m be the inclusion of $\ker h$ into A^2 (thus $\ker h$ can be interpreted as a T-algebra). Then F is said to *preserves kernels*, iff

$$< \Pi \circ F\, m[F\,(\ker h)] > = \ker F\, h \cap \Pi[F\, A^2]$$

where $< \_ >$ denotes congruence generation.

Theorem 6.3.2 With respect to specifiable PADTs, our hierarchy has the following properties:

6.3. PROPERTIES OF PARAMETERIZED ADTS

	Institution	Properties of specifiable functors F	Separating example (constructed in section 6.4)
1	$(=)$	Fh is full	
2	$RP(R \stackrel{f}{=})$	F preserves kernels	Making a relation reflexive
3	$RP(=\Rightarrow=)(R \stackrel{f}{=})$	F preserves full surjective homomorphisms	Factorization of a function over the image
4	$RP(R \stackrel{f}{=} \Rightarrow R \stackrel{f}{=})$	F preserves surjective homomorphisms	Making a binary relation transitive
5	$RP(R \stackrel{e}{=} \Rightarrow R \stackrel{e}{=})$		Generating the paths in a multigraph

Proof: (1) Level 5 versus level 4: In $RP(R \stackrel{f}{=} \Rightarrow R \stackrel{f}{=})$, specifiable functors preserve surjectivity.

Since forgetful functors obviously preserve surjectivity, w. l. o. g. we can assume that F is a free $T \stackrel{\theta}{\longrightarrow} T1$-PADT in $RP(R \stackrel{f}{=} \Rightarrow R \stackrel{f}{=})$. Let $h\colon A \longrightarrow B$ be a surjective T-homomorphism. We want to show that Fh is surjective as well.

This is done in two steps. First we represent the free construction as a quotient of a term algebra $T_{\Sigma 1}(A)$ over the parameter A. Then Fh acts as inductive extension of h to terms, which is denoted by $T_{\Sigma 1}(h)$, with $\Sigma 1 = sign(T1)$. In a second step, we "factor out" the parameter A by representing a term from $T_{\Sigma 1}(A)$ by a term from the term algebra $T_{\Sigma 1}(Y)$ over a fixed variable set Y, together with a valuation $\nu\colon Y \longrightarrow A$, which is used to replace the leaves of the term. With this representation, $T_{\Sigma 1}(h)$ acts by composing a valuation with h, and this is surjective because h is surjective.

More formally, let $\mu^{graph}\colon RP(R \stackrel{f}{=} \Rightarrow R \stackrel{f}{=}) \longrightarrow R(R =\Rightarrow R =)$ the obvious extension of the representation introduced in the second half of the proof of proposition 4.8.1 (relation symbols are just left unchanged). Let $\Sigma 1$ be the signature of $\Phi^{graph} T1$ and let F' be the free $\Phi^{graph} T \xrightarrow{\Phi^{graph} \theta} \Phi^{graph} T1$-PADT. Since $\Sigma 1$ contains only total operation symbols, the construction of fact 3.10.1 has the property that $H_{T1}(A) = T_{\Sigma 1}(A)$. Thus there is a quotient $T_{\Sigma 1}(\beta_T^{graph^{-1}} A) \xrightarrow{q_A} F' \beta_T^{graph^{-1}} A$. Because β_T^{graph} and β_{T1}^{graph} are the identity on carrier sets, we can represent $|FA|$ as quotient (full surjection) $|T_{\Sigma 1}(A)| \xrightarrow{q_A} |FA|$. Similarly, we get a quotient $|T_{\Sigma 1}(B)| \xrightarrow{q_B} |FB|$.

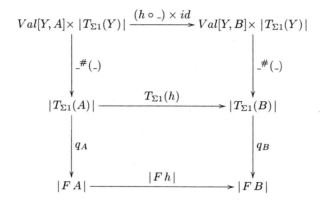

Since the lower rectangle commutes, surjectivity of Fh now follows from surjectivity of $T_{\Sigma 1}(h)$. To show the latter, fix some infinite S-sorted variable system Y, where S are the sorts of T. We represent $|T_{\Sigma 1}(A)|$ by $Val[Y, A] \times T_{\Sigma 1}(Y)$. A pair $(\nu: Y \longrightarrow A, t)$ is mapped to $\nu^\#(t) \in |T_{\Sigma 1}(A)|$. Since each term in $|T_{\Sigma 1}(A)|$ only uses finitely many elements of A, this assignment is surjective. A similar representation holds for B. By commutativity of the upper diagram, surjectivity of $T_{\Sigma 1}(h)$ now follows from surjectivity of $(h \circ \_) \times id$. To prove the latter, let $\nu: Y \longrightarrow B$ some valuation. By surjectivity of h, there is some factorization $Y \xrightarrow{\nu} B = Y \xrightarrow{\lambda} A \xrightarrow{h} B$. Thus, $h \circ \_$ is surjective.

(2) Level 4 versus level 3: In $RP(=\Rightarrow=)(R \stackrel{f}{=})$, specifiable functors preserve full surjective homomorphisms. By observing that $\beta^{(3.10)\circ(3.9)^{-1}}$ (defined in section 5.4) exactly maps full surjective homomorphisms to surjective homomorphisms, we can use (1). More exactly, full surjectivity of $h: A \longrightarrow B$ means surjectivity and

$$h[graph(pop_A)] = graph(pop_B) \quad (pop: s_1, \ldots, s_n \longrightarrow s \in POP)$$
$$h[R_A] = R_B \quad (R: s_1, \ldots, s_n \in REL)$$

These two are equivalent to surjectivity of $(\beta^{(3.10)\circ(3.9)^{-1}}_T h)_{s^{pop}}$ $(pop \in POP)$ and $(\beta^{(3.10)\circ(3.9)^{-1}}_T h)_{s^R}$ $(R \in REL)$, respectively.

Assume that F is a free $T \xrightarrow{\theta} T1\text{-PADT}$ (now in $RP(=\Rightarrow=)(R \stackrel{f}{=})$) and $h: A \longrightarrow B$ is a full surjective homomorphism. By the above equivalence, $\beta^{(3.10)\circ(3.9)^{-1}}_T h$ is surjective as well. Let F' be the free $\Phi^{(3.10)\circ(3.9)} T \xrightarrow{\Phi^{(3.10)\circ(3.9)} \theta} \Phi^{(3.10)\circ(3.9)} T1\text{-PADT}$. By (1), we have that $F' \beta^{(3.10)\circ(3.9)^{-1}}_T h \cong \beta^{(3.10)\circ(3.9)^{-1}}_{T1} Fh$ is surjective. By the above equivalence, Fh is full and surjective.

6.3. Properties of parameterized ADTs

(3) Level 3 versus level 2: In $RP(R \stackrel{f}{=})$, specifiable functors preserve kernels. With a similar argument as in (1), we can use μ^{graph} to restrict ourselves to $R(R =)$. Now

$$\begin{array}{ccc} T_{\Sigma 1}(A) & \xrightarrow{T_{\Sigma 1}(h)} & T_{\Sigma 1}(B) \\ \downarrow q_A & & \downarrow q_B \\ FA & \xrightarrow{Fh} & FB \end{array}$$

commutes in $\mathbf{Mod}^{R(R=)}(\Sigma 1)$. Then we have

Lemma 6.3.3 For surjective T-homomorphisms h,

$$\ker(q_B \circ T_{\Sigma 1}(h)) = < \ker q_A \cup \ker T_{\Sigma 1}(h) >$$

Proof: "$\supseteq$": This follows from

$$\ker q_A \subseteq \ker(F h \circ q_A) = \ker(q_B \circ T_{\Sigma 1}(h)) \text{ and}$$

$$\ker T_{\Sigma 1}(h) \subseteq \ker(q_B \circ T_{\Sigma 1}(h))$$

"$\subseteq$": In $(=)$ and also in $R(R =)$, we have that logical consequence is expressible by congruence generation, see [EM85]. The equation for $\ker q_A$ from the proof of fact 3.10.1 can thus be rewritten as

$$\ker q_A = < \{ (\nu^\#(t_1), \nu^\#(t_2)) \mid \nu: X \longrightarrow T_{\Sigma 1}(A), (X : t_1 = t_2) \in \Gamma 1(A) \} >$$

where $\Gamma 1(A)$ is defined in the above mentioned proof. Thus $(t_1, t_2) \in \ker(q_B \circ T_{\Sigma 1}(h))$ implies $(T_{\Sigma 1}(h)(t_1), T_{\Sigma 1}(h)(t_2)) \in \ker q_B$, which implies that there is some $\nu: X \longrightarrow T_{\Sigma 1}(B)$ and some $(X : u_1 = u_2) \in \Gamma 1(B)$ with $\nu^\#(u_1) = T_{\Sigma 1}(h)(t_1)$ and $\nu^\#(u_2) = T_{\Sigma 1}(h)(t_2)$. By surjectivity of h (and thus, by (1), of $T_{\Sigma 1}(h)$), we can factor $X \xrightarrow{\nu} T_{\Sigma 1}(B)$ as $X \xrightarrow{\lambda} T_{\Sigma 1}(A) \xrightarrow{T_{\Sigma 1}(h)} T_{\Sigma 1}(B)$, so that $\nu^\# = T_{\Sigma 1}(h) \circ \lambda^\#$. Fix $(X : v_1 = v_2) \in \Gamma 1(A)$ with $u_i = T_{\Sigma 1}(h)(v_i)$ $(i = 1, 2)$. Now $T_{\Sigma 1}(h)(t_i) = \nu^\#(u_i) = T_{\Sigma 1}(h)(\lambda^\#(v_i))$ $(i = 1, 2)$. Thus $(t_1, \lambda^\#(v_1)) \in \ker T_{\Sigma 1}(h), (\lambda^\#(v_1), \lambda^\#(v_2)) \in \ker q_A$ and $(\lambda^\#(v_2), t_2) \in \ker T_{\Sigma 1}(h)$, so altogether $(t_1, t_2) \in < \ker q_A \cup \ker T_{\Sigma 1}(h) >$. □

Lemma 6.3.3 is now the key for relating free construction and term algebra in the proof of

$$< \Pi \circ F m[F \ker h] > = \ker F h \cap \Pi[F A^2]$$

Let A^2 be the product of A with itself, with projections $\pi_1, \pi_2: A^2 \longrightarrow A$. Let $\Pi: F A^2 \longrightarrow (F A)^2 := < F \pi_1, F \pi_2 >$, that is, Π is the unique $T1$-

homomorphism with

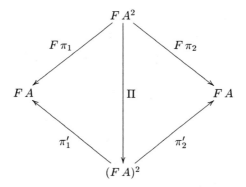

commuting, where $\pi'_1, \pi'_2 \colon (FA)^2 \longrightarrow FA$ are the projections of $(FA)^2$ onto its factors.

Let now (η, F) be a free $T \xrightarrow{\theta} T1\text{-PADT}$ and $h \colon A \longrightarrow B$ be a surjective T-homomorphism. Let m be the inclusion of $\ker h$ into A^2 (thus $\ker h$ can be interpreted as a T-algebra).

Let $\Pi' :=< T_{\Sigma 1}\pi_1, T_{\Sigma 1}\pi_2 > \colon T_{\Sigma 1} A^2 \longrightarrow (T_{\Sigma 1} A)^2$ defined in a similar way as above.

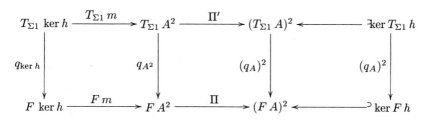

Lemma 6.3.4   1. For a binary relation $R \subseteq (T_{\Sigma 1} A)^2$, we have $(q_A)^2[<R>] \subseteq <(q_A)^2[R]>$

2. $\Pi \circ Fm[F \ker h] = (q_A)^2[\ker T_{\Sigma 1} h \cap \Pi'[T_{\Sigma 1} A^2]]$

3. $\ker F h \cap \Pi[F A^2] = (q_A)^2[\ker(q_B \circ T_{\Sigma 1} h)] \cap (q_A)^2 \circ \Pi'[T_{\Sigma 1} A^2]$

Proof:

1. Proof by induction over the definition of $<R>$.

   (1) Clearly, $(q_A)^2[R] \subseteq <(q_A)^2[R]>$

6.3. PROPERTIES OF PARAMETERIZED ADTS

(2) $(q_A)^2[\Delta\, T_{\Sigma 1}(A)] \subseteq\; <(q_A)^2[R]>$

(3) $(t, u) \in <R>$ by inductive hypothesis implies $(q_A(t), q_A(u)) \in <(q_A)^2[R]>$, which implies $(q_A(u), q_A(t)) \in <(q_A)^2[R]>$

(4) Transitivity and compatibility with operations are treated similarly.

2. We have $\Pi'[T_{\Sigma 1}\, \ker h] = \{((\pi_1 \circ \nu)^\#(t), (\pi_2 \circ \nu)^\#(t)) \mid X$ an S-sorted variable system, $\nu\colon X \longrightarrow A^2$, $t \in T_{\Sigma 1}(X)$, $h \circ \pi_1 \circ \nu = h \circ \pi_2 \circ \nu\} = \ker T_{\Sigma 1}\, h \cap \Pi'[T_{\Sigma 1}\, A^2]$. Therefore,

$\Pi \circ F\, m[F\, \ker h]$
$= \Pi \circ F\, m \circ q_{\ker h}[T_{\Sigma 1}\, \ker h]$
$= \Pi \circ q_{A^2} \circ T_{\Sigma 1}\, m[T_{\Sigma 1}\, \ker h]$
$= (q_A)^2 \circ \Pi'[T_{\Sigma 1}\, \ker h]$
$= (q_A)^2[\ker T_{\Sigma 1}\, h \cap \Pi'[T_{\Sigma 1}\, A^2]]$

3. $\quad \ker F\, h \cap \Pi[F\, A^2]$
$= (q_A)^2[\ker(F\, h \circ q_A)] \cap (q_A)^2 \circ \Pi'[T_{\Sigma 1}\, A^2]$
$= (q_A)^2[\ker(q_B \circ T_{\Sigma 1}\, h)] \cap (q_A)^2 \circ \Pi'[T_{\Sigma 1}\, A^2]$ □

Now we are ready to prove

$$<\Pi \circ F\, m[F\, \ker h]> = \ker F\, h \cap \Pi[F\, A^2]$$

"$\subseteq$":

$\Pi \circ F\, m[F\, \ker h]$
$= (q_A)^2[\ker T_{\Sigma 1}\, h \cap \Pi'[T_{\Sigma 1}\, A^2]]$ (by 2.)
$= (q_A)^2[\ker T_{\Sigma 1}\, h] \cap (q_A)^2 \circ \Pi'[T_{\Sigma 1}\, A^2]$
$\subseteq (q_A)^2[\ker(q_B \circ T_{\Sigma 1}\, h)] \cap (q_A)^2 \circ \Pi'[T_{\Sigma 1}\, A^2]$
$= \ker F\, h \cap \Pi[F\, A^2]$ (by 3.)

Since the latter is a congruence, the inclusion holds for $<\Pi \circ F\, m[F\, \ker h]>$ as well.

"$\supseteq$":

$\ker F\, h \cap \Pi[F\, A^2]$
$= (q_A)^2[\ker(q_B \circ T_{\Sigma 1}\, h)] \cap (q_A)^2 \circ \Pi'[T_{\Sigma 1}\, A^2]$ (by 3.)
$= (q_A)^2[<\ker q_A \cup \ker T_{\Sigma 1}\, h>] \cap (q_A)^2 \circ \Pi'[T_{\Sigma 1}\, A^2]$ (by Lemma 6.3.3)
$\subseteq <(q_A)^2[\ker q_A] \cup (q_A)^2[\ker T_{\Sigma 1}\, h]> \cap (q_A)^2 \circ \Pi'[T_{\Sigma 1}\, A^2]$ (by 1.)
$= <\Delta\, F\, A \cup (q_A)^2[\ker T_{\Sigma 1}\, h]> \cap (q_A)^2 \circ \Pi'[T_{\Sigma 1}\, A^2]$
$= <(q_A)^2[\ker T_{\Sigma 1}\, h]> \cap (q_A)^2 \circ \Pi'[T_{\Sigma 1}\, A^2]$
$= <(q_A)^2[\ker T_{\Sigma 1}\, h \cap \Pi'[T_{\Sigma 1}\, A^2]]>$
$= <\Pi \circ F\, m[F\, \ker h]>$ (by 2.)

(4) Level 2 versus level 1:

In (=), all homomorphisms are full (this is in fact a property of the model classes). □

6.4 Example PADTs and a hierarchy theorem

In this section, we examine some example PADTs that are simple but useful to illustrate the results of the preceding sections. In particular, we look for example PADTs separating the different methods.

The first example can be specified at level 5 in $P(\stackrel{e}{=}\Rightarrow\stackrel{e}{=})$:

Example 6.4.1 Finding the paths over a graph is specifiable at level 5 with $P(\stackrel{e}{=}\Rightarrow\stackrel{e}{=})$, but not at level 4 resp. with $RP(R \stackrel{f}{\Longrightarrow} R \stackrel{f}{=})$. Let **(GRAPH,PATHS)** be the following parameterized theory:

spec GRAPH =
  sorts *nodes, edges*
  ops *source, target*: *edges* ⟶ *nodes*

spec PATHS[GRAPH] =
  ops *conc*: *edges* × *edges* ⟶? *edges*
  $\forall x, y$: *nodes* . $source(y) = target(x) \Longleftrightarrow conc(x,y) \stackrel{e}{=} conc(x,y)$
  $\forall x, y$: *nodes* . $source(y) = target(x) \Longrightarrow target(conc(x,y)) \stackrel{e}{=} target(y)$
  $\forall x, y$: *nodes* . $source(y) = target(x) \Longrightarrow source(conc(x,y)) \stackrel{e}{=} source(x)$

Let (η, F) be the free **(GRAPH,PATHS)**-PADT, and let G_1 and G_2 **GRAPH**-algebras representing the following graphs:

6.4. Example PADTs and a hierarchy theorem

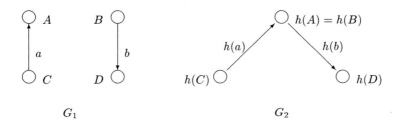

There is a closed surjective **GRAPH**-homomorphism $h: G_1 \longrightarrow G_2$ indicated in the picture. Now the **PATHS**-algebras $G_3 = F\,G_1$ and $G_4 = F\,G_2$ can be computed. They look like:

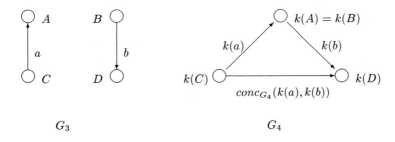

The homomorphism $k = F\,h$ acts similarly to h. The only difference is that the edge $conc_{G_4}(k(a), k(b))$ has no pre-image. Thus, k is not surjective, and F does not preserve surjectivity. By theorem 6.3.2, (η, F) cannot be specifiable at level 4 resp. with $RP(R \stackrel{f}{=}\!\!\Rightarrow R \stackrel{f}{=})$, although it is specifiable at level 5 with $P(\stackrel{e}{=}\!\!\Rightarrow \stackrel{e}{=})$. □

Example 6.4.2 Transitive closure of a relation is specifiable at level 4 with $R(R =\!\!\Rightarrow R =)$, but not at level 3 resp. with $RP(=\!\!\Rightarrow =)(R \stackrel{f}{=})$.

spec BIN_REL =
  sorts *elem*
  preds $R : elem \times elem$

spec TRANS_REL[BIN_REL] =
$\forall x, y, z : elem . \ R(x,y) \land R(y,z) \Longrightarrow R(x,z)$

Now let (η, F) be the free **(BIN_REL,TRANS_REL)**-PADT. Interpret the pictures above as **BIN_REL**-algebras, with an arrow from X to Y meaning that $R(X,Y)$. Then $h: G_1 \longrightarrow G_2$ is surjective and full (even closed), but $k = Fh$ is not full.

Thus, by theorem 6.3.2, (η, F) cannot be specifiable with at level 3, although it is specifiable at level 4. □

Example 6.4.3 Factorization of a function over the image is specifiable at level 3 with $(=\!\Rightarrow\!=)$, but not at level 2 resp. with $RP(R \stackrel{f}{=})$.

spec FUNC =
 sorts *source, target*
 ops f: *source* $\longrightarrow$ *target*

spec FACTOR[FUNC] =
 sorts *image*
 ops $surj$: *source* $\longrightarrow$ *image*
  inj: *image* $\longrightarrow$ *target*
 $\forall x$: *source* . $inj(surj(x)) = f(x)$
 $\forall x, y$: *source* . $f(x) = f(y) \Longrightarrow surj(x) = surj(y)$

Let (η, F) be the free **(FUNC,FACTOR)**-PADT. F takes a function f and yields its factorization over the image, that is, f is expressed as a surjection followed by an injection.

Let $h: A \longrightarrow B$ be the following **FUNC**-homomorphism:

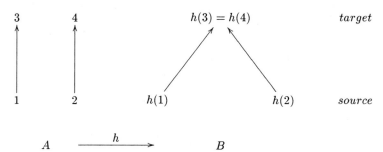

6.4. Example PADTs and a hierarchy theorem

The graph of f is indicated by arrows in the figure. The kernel of h is the following **FUNC**-algebra:

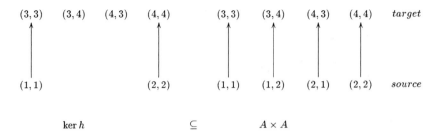

If we apply F to the whole thing, we get

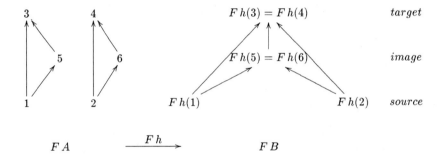

Now $FA^2 \cong (FA)^2$, we even can take the identity here as isomorphism. $F(\ker h)$ has the form

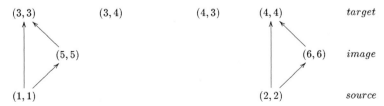

$F \ker h$

while $\ker(Fh)$ has the form

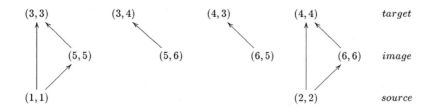

ker $F h$

Since $\Pi : F A^2 \longrightarrow (FA)^2$ is the identity and $F(\ker h)$ is already a congruence, we have $< \Pi \circ F m[F(\ker h)] > = F \ker h \neq \ker(F h) = \ker(F h) \cap \Pi[F A^2]$. Thus, by theorem 6.3.2, F cannot be specified at level 2 resp. with $RP(R \stackrel{f}{=})$.
□

Example 6.4.4 Making a relation reflexive is specifiable at level 2 with $RP(R \stackrel{f}{=})$, but not at level 1 resp. with $(=)$. Making a relation reflexive can be specified as follows:

spec BIN_REL =
  sorts s
  preds $R : s \times s$

spec REFLEXIVE[BIN_REL] =
$\forall x : s .\ R(x, x)$

Now this is not specifiable within $(=)$ since there are no relations in $(=)$. (This is a bit too trivial, of course, and was the motivation for equipping all the other institutions with relations and partiality. But by Corollary 6.1.11, relations and partiality are not encodable in $(=)$.)

Altogether, theorem 6.3.2 together with examples 6.4.1, 6.4.2, 6.4.3 and 6.4.4 lead to the following hierarchy theorem.

Theorem 6.4.5 (Hierarchy Theorem) The institutions $RP(R \stackrel{e}{=} \Rightarrow R \stackrel{e}{=})$, $RP(R \stackrel{f}{=} \Rightarrow R \stackrel{f}{=})$, $RP(=\Rightarrow=)(R \stackrel{f}{=})$, $RP(R \stackrel{f}{=})$ and $(=)$ form a hierarchy of strictly decreasing expressive power with respect to the specifiability of PADTs.
□

6.5 Locating bounded stacks in the hierarchy

Consider the following requirement:

spec BOUNDED_ELEM = NAT and BOOL then
  sorts *elem*
  ops *inbound*: *nat* ⟶ *bool*
  $\forall n\!:\!nat$. $inbound(succ(n)) = true \implies inbound(n) = true$
  $\forall n\!:\!nat$. $inbound(n) = false \implies inbound(succ(n)) = false$

spec BOUNDED_STACK[BOUNDED_ELEM] given NAT and BOOL =
  sorts *stack*
  ops *mt* : *stack*
    push: *elem stack* ⟶? *stack*
    pop: *stack* ⟶? *stack*
    top: *stack* ⟶? *stack*

Let C be the subcategory of $\mathbf{Mod}^{RP(=\Rightarrow=)(R\stackrel{f}{=})}$ **BOUNDED_ELEM** which contains all models with **NAT**-reduct isomorphic to the natural numbers and **BOOL**-reduct isomorphic to the standard booleans. Our aim is to specify the persistent functor **BSTACK**: $C \longrightarrow \mathbf{Mod}^{RP(=\Rightarrow=)(R\stackrel{f}{=})}$ **BOUNDED_STACK** such that **BSTACK**$(A)_{stack}$ contains all stacks over A_{elem} with heights n for which $inbound_A(succ^n_A(0_A)) = true_A$, with *pop*, *push* and *top* the usual (partial!) stack operations.

Now in $P(\stackrel{e}{=}\Rightarrow\stackrel{e}{=})$, this is easy:

spec BOUNDED_STACK_BODY1[BOUNDED_ELEM]
    given NAT and BOOL =
ops *height*: *stack* ⟶ *nat*
. $height(mt) = 0$
$\forall m\!:\!elem, s\!:\!stack$. $inbound(succ(height(s))) = true \iff$
    $push(m, s) \stackrel{e}{=} push(m, s)$
$\forall m\!:\!elem, s\!:\!stack$. $push(m, s) \stackrel{e}{=} push(m, s) \implies$
    $height(push(m, s)) \stackrel{e}{=} succ(height(s))$
$\forall m\!:\!elem, s\!:\!stack$. $push(m, s) \stackrel{e}{=} push(m, s) \implies pop(push(m, s)) \stackrel{e}{=} s$
$\forall m\!:\!elem, s\!:\!stack$. $push(m, s) \stackrel{e}{=} push(m, s) \implies top(push(m, s)) \stackrel{e}{=} m$

The free **BOUNDED_ELEM**↪**BOUNDED_STACK_BODY1**-functor composed with the forgetful functor from **Mod BOUNDED_STACK_BODY1** to **Mod BOUNDED_STACK**, when restricted to C, is equal to **BSTACK**. The free functor does not preserve surjectivity (consider homomorphisms into the terminal model) and is thus not specifiable in $RP(R \stackrel{f}{=}\!\!\Rightarrow R \stackrel{f}{=})$ by theorem 6.3.2.

So we have to look for a different way to specify **BSTACK** in $RP(R \stackrel{f}{=}\!\!\Rightarrow R \stackrel{f}{=})$. Instead of generating bounded stacks by a partial operation (which is impossible in $RP(R \stackrel{f}{=}\!\!\Rightarrow R \stackrel{f}{=})$), we generate them by a total operation, factor out those stacks that are too high and then specify the partial push as a derived operation.

spec BOUNDED_STACK_BODY2[BOUNDED_ELEM]
       given NAT and BOOL =
ops $push1$: $elem\ stack \longrightarrow stack$
     $height$: $stack \longrightarrow nat$
. $height(mt) = 0$
$\forall m$: $elem, s$: $stack$. $inbound(height(s)) = true \Longrightarrow$
       $height(push1(m,s)) = succ(height(s))$
$\forall m$: $elem, s$: $stack$. $inbound(height(s)) = false \Longrightarrow push1(m,s) = s$
$\forall m$: $elem, s$: $stack$. $inbound(height(s)) = true \Longrightarrow push(m,s) \stackrel{e}{=} push1(m,s)$
$\forall m$: $elem, s$: $stack$. $inbound(height(s)) = true \Longrightarrow pop(push1(m,s)) \stackrel{e}{=} s$
$\forall m$: $elem, s$: $stack$. $inbound(height(s)) = true \Longrightarrow top(push1(m,s)) \stackrel{e}{=} m$

What about $RP(=\!\!\Rightarrow=)(R \stackrel{f}{=})$? Consider the **BOUNDED_ELEM**-algebras $A = (I\!N, \{false, true\}, \{*\}, inbound_A)$ with $inbound_A(n) = false$ and $B = (I\!N, \{false, true\}, \{*\}, inbound_B)$ with $inbound_B(n) = true$. Let $A \stackrel{h}{\longrightarrow} 1$ and $B \stackrel{k}{\longrightarrow} 1$ be the unique homomorphisms into the terminal object. They are full and surjective. Since **BSTACK** B are unbounded stacks over $\{*\}$, $push_{\mathbf{BSTACK}\,B}(*, mt_{\mathbf{BSTACK}\,B})$ is defined. Therefore $push_{\mathbf{BSTACK}\,1}(\mathbf{BSTACK}\,k(*), \mathbf{BSTACK}\,k(mt_{\mathbf{BSTACK}\,B})) = push_{\mathbf{BSTACK}\,1}(*, mt_{\mathbf{BSTACK}\,1})$ is defined as well.

On the other hand, **BSTACK** A has a totally undefined $push$. In particular, $push_{\mathbf{BSTACK}\,A}(*, mt_{\mathbf{BSTACK}\,A})$ is undefined, but

$$push_{\mathbf{BSTACK}\,1}(*, mt_{\mathbf{BSTACK}\,1})$$
$$= push_{\mathbf{BSTACK}\,1}(\mathbf{BSTACK}\,h(*), \mathbf{BSTACK}\,h(mt_{\mathbf{BSTACK}\,A}))$$

is defined. This means that **BSTACK** h cannot be full. This implies **BSTACK** does not preserve full surjectivity, and by theorem 6.3.2, in $RP(=\!\!\Rightarrow=)(R \stackrel{f}{=})$, **BSTACK** cannot be specified!

6.6 Summary

We have proved a hierarchy theorem concerning widely-used institutions, ranging from equational logic $((=))$ to Relational Partial Conditional Existence-

6.6. Summary

Equation Logic ($RP(R \stackrel{e}{=}\!\!\Rightarrow R \stackrel{e}{=})$). While the institutions do not differ with respect to initial semantics, they do differ in their properties of model categories and of PADTs.

I argue that differences in the hierarchy are *not* caused by the availability of partiality and/or relations. This is shown by introducing institutions having those features that are (weakly) equivalent in expressiveness to the widely-used institutions that not (all) have these features. Then, the real cause of the differences becomes visible: the degree of conditionality in the axioms we may use.

In the most expressive institutions $RP(R \stackrel{e}{=}\!\!\Rightarrow R \stackrel{e}{=})$ and $P(\stackrel{e}{=}\!\!\Rightarrow\!\stackrel{e}{=})$, we have *conditional generation of data*. This corresponds to availability of conditional axioms with conclusion $pop(t_1, \ldots, t_n) \stackrel{e}{=} pop(t_1, \ldots, t_n)$, i.e. a conclusion that may generate data in the free construction. On the other hand side, in $RP(R \stackrel{f}{=}\!\!\Rightarrow R \stackrel{f}{=})$ and $R(R =\!\!\Rightarrow R =)$, we only have unconditional generation of data in the free construction. This corresponds to the property of PADTs to preserve surjective homomorphisms.

However, in $RP(R \stackrel{f}{=}\!\!\Rightarrow R \stackrel{f}{=})$ and $R(R =\!\!\Rightarrow R =)$, we can *conditionally generate members of relations or of graphs of partial operations* using conclusions of the form $R(t_1, \ldots, t_n)$ or $pop(t_1, \ldots, t_n) \stackrel{e}{=} t_0$. On the other hand side, in $RP(=\!\!\Rightarrow=)(R \stackrel{f}{=})$ and $(=\!\!\Rightarrow=)$, this is not possible, and PADTs here preserve full surjective homomorphisms.

Now in $RP(=\!\!\Rightarrow=)(R \stackrel{f}{=})$ and $(=\!\!\Rightarrow=)$, we have the same thing with *conditional generation of kernel members*, which is not possible in $RP(R \stackrel{f}{=})$ and $(=)$, where kernels are preserved by PADTs.

This is illustrated by the following table:

Institution	We have conditional generation of	This corresponds to availability of conditional axioms with conclusion	But we have only unconditional generation of	This corresponds to PADT property
$(=)$, $RP(R \stackrel{f}{=})$			kernel members (equations)	F preserves kernels
$(=\Rightarrow=)$, $P(=\Rightarrow=)(\stackrel{f}{=})$	kernel members (equations)	$e_1 \wedge \ldots \wedge e_n \Longrightarrow t_1 = t_2$	relation/ graph members	F preserves full surjective homomorphisms
$R(R \Longrightarrow R=)$, $RP(R \stackrel{f}{=} \Longrightarrow R \stackrel{f}{=})$	relation/ graph members	$e_1 \wedge \ldots \wedge e_n \Longrightarrow R(t_1, \ldots, t_n)$	data	F preserves surjective homomorphisms
$P(\stackrel{e}{=}\Rightarrow\stackrel{e}{=})$, $RP(R \stackrel{e}{=} \Longrightarrow R \stackrel{e}{=})$	data	$e_1 \wedge \ldots \wedge e_n \Longrightarrow pop(t_1, \ldots, t_n) \stackrel{e}{=} pop(t_1, \ldots, t_n)$		

It might be interesting to have purely categorical separating properties of PADTs as well, because they would be institution independent. Indeed, preservation of kernels is such a property. For the other cases, it is impossible that such a property be found (at least if the property is preserved by categorical retractive representations) due to the results of chapter 7.

Chapter 7

Equivalence and Difference of Institutions: Simulating Horn Clause Logic With Based Algebras

> "This choice of semantical simplicity, together with the fact that no restriction is put on the typing relation, resulted in the following, somewhat surprising outcome: formally, ET logic can be viewed as Horn logic with equality and one (binary) predicate, viz. type assignment. This fact renders the parsimony of this logic very well. However, the obvious conclusion that ET logic *is* a special case of Horn logic can easily prove misleading. Types and predicates are very different logical devices, although either has sufficient expressive power to 'encode' the other. The differences, which arise from the intuition that underlies these two formal devices, are reflected in their 'natural' use..." *V. Manca, A. Salibra and G. Scollo* [MSS90]

The slogan of this chapter is that this "natural use" and "different logical devices" can be technically captured by the notion of categorical retractive representation. That is, "natural use" refers to a second logical system, which is encoded via a categorical retractive representation into ET logic (or approaches of similar structure, like unified algebras [Mos89] or the *PART* construction on based algebras [Kre87]). Moreover, in some cases it is possible to show that the essential difference between the two logical systems is captured by the categorical retractive representation as well.

The two logical systems occurring in such a situation typically are not equivalent in expressiveness in the sense of chapter 4, but are still relatively close to each other. Thus the notion of embedding is too narrow here and should be weakened. Therefore, we introduce a notion of categorical retractive representation, which captures the essential properties of many representations of partiality in total logical systems. Finally, this is used to develop a measure of equivalence and difference between specification frames (which, by the functor $\mathbf{Th_0}\colon \mathbf{Inst} \longrightarrow \mathbf{SpecFram}$, also applies for institutions).

Parts of this chapter have been published in [KM95].

7.1 Conditional Equational Logic and Based Algebras

Compared with $R(R \Longrightarrow R =)$, in $(\Longrightarrow=)$ we loose the possibility of representing partial operations as graphs. But there is another possibility: Kreowski [Kre87] has introduced based algebras as a device for different purposes. In particular, based algebras can be used to gain partiality while staying in a total approach (without relations).

The institution $B(\Longrightarrow=)$ of based algebras and based theories has as signatures triples of the form $(\Sigma_B, \Sigma_C, \sigma)$, where Σ_B and Σ_C are $(\Longrightarrow=)$-signatures ("B" indicates the base part, "C" the target part), and $\sigma\colon \Sigma_B \longrightarrow \Sigma_C$ is a $(\Longrightarrow=)$-signature morphism. $B(\Longrightarrow=)$-signature morphisms $\theta\colon (\Sigma_B, \Sigma_C, \sigma) \longrightarrow (\Sigma'_B, \Sigma'_C, \sigma')$ are pairs $(\theta_B\colon \Sigma_B \longrightarrow \Sigma'_B, \theta_C\colon \Sigma_C \longrightarrow \Sigma'_C)$, such that $\theta_C \circ \sigma = \sigma' \circ \theta_B$ in $(\Longrightarrow=)$.
A $(\Sigma_B, \Sigma_C, \sigma)$-algebra is a triple (B, C, b), where B is a Σ_B-algebra, C is a Σ_C-algebra and $b\colon B \longrightarrow C\restriction_\sigma$ is a Σ_B-homomorphism. Homomorphisms $h = (h_B, h_C)\colon (B, C, b) \longrightarrow (B', C', b')$ consist of a Σ_B-homomorphism $h_B\colon B \longrightarrow B'$ and a Σ_C-homomorphism $h_C\colon C \longrightarrow C'$ such that

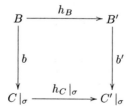

commutes.
In other words: $\mathbf{Mod}^{B(\Longrightarrow=)}(\Sigma_B, \Sigma_C, \sigma)$ is the comma category $(\mathbf{Mod}^{(\Longrightarrow=)}\Sigma_B, \mathbf{Mod}^{(\Longrightarrow=)}\sigma)$. Reducts are defined component wise.

7.1. Conditional Equational Logic and Based Algebras

If $(\Sigma_B, \Sigma_C, \sigma)$ is a $B(\!=\!\Rightarrow\!=\!)$-signature, let Σ_{BC} be $\Sigma_B \dot\cup \Sigma_C$ plus operations

$$b_s: inj_B(s) \longrightarrow inj_C(\sigma(s)) \text{ for } s \in S(\Sigma_B),$$

where $inj_B: \Sigma_B \longrightarrow \Sigma_B \dot\cup \Sigma_C$ and $inj_C: \Sigma_C \longrightarrow \Sigma_B \dot\cup \Sigma_C$ are the injections. Then a $(\Sigma_B, \Sigma_C, \sigma)$-sentence is a Σ_{BC}-sentence in $(\!=\!\Rightarrow\!=\!)$.

If (B, C, b) is a $(\Sigma_B, \Sigma_C, \sigma)$-algebra, let $\beta^{flatten^{-1}}(B, C, b)$ be the Σ_{BC}-algebra A with $A\,|_{inj_B} = B$, $A\,|_{inj_C} = C$ and $b_{s,A} = b_s$. Now satisfaction is defined as follows:

$$(B, C, b) \models^{B(=\Rightarrow=)}_{(\Sigma_B, \Sigma_C, \sigma)} \varphi \iff A \models^{(=\Rightarrow=)}_{\Sigma_{BC}} \varphi$$

By abuse of terminology, we speak of valuations into (B, C, b) when we in fact mean valuations into $\beta^{flatten^{-1}}(B, C, b)$.

Proposition 7.1.1 $(\!=\!\Rightarrow\!=\!)$ and $B(\!=\!\Rightarrow\!=\!)$ are equivalent in expressiveness.

Proof: Define $\mu^{embed}: (\!=\!\Rightarrow\!=\!) \longrightarrow B(\!=\!\Rightarrow\!=\!)$ by

- $\Phi^{embed}(\Sigma) = (\emptyset, \Sigma, \emptyset \hookrightarrow \Sigma)$
- $\alpha^{embed}_\Sigma(\varphi) = \varphi$ (note that $\Sigma_{BC} = \Sigma$ here)
- $\beta^{embed}_T(B, C, b) = C$
- $\beta^{embed}_T(h_B, h_C) = h_C$

Actually, Φ^{embed} preserves colimits since it is left-adjoint as in proposition 4.8.1.
Define $\mu^{flatten}: B(\!=\!\Rightarrow\!=\!) \longrightarrow (\!=\!\Rightarrow\!=\!)$ by

- $\Phi^{flatten}(\Sigma_B, \Sigma_C, \sigma)$ is Σ_{BC} with axioms
  $(\forall x_1: s_1, \dots, x_n: s_n .$
  $\quad b_s(op(x_1, \dots, x_n)) = \sigma(op)(b_{s_1}(x_1), \dots, b_{s_n}(x_n)))$
  for $op: s_1, \dots, s_n \longrightarrow s \in OP(\Sigma_B)$
- $\alpha^{flatten}_{(\Sigma_B, \Sigma_C, \sigma)}(\varphi) = \varphi$
- $\beta^{flatten}_{(\Sigma_B, \Sigma_C, \sigma)}(A) = (A\,|_{\Sigma_B \hookrightarrow \Sigma_{BC}}, A\,|_{\Sigma_C \hookrightarrow \Sigma_{BC}}, (b_{s,A})_{s \in S(\Sigma_B)})$
- $\beta^{flatten}_{(\Sigma_B, \Sigma_C, \sigma)}(h) = (h\,|_{\Sigma_B \hookrightarrow \Sigma_{BC}}, h\,|_{\Sigma_C \hookrightarrow \Sigma_{BC}})$

$\Phi^{flatten}$ also preserves colimits. This basically follows from coproducts and coequalizers being constructed component wise in $B(\!=\!\Rightarrow\!=\!)$. To see that coequalizers are constructed component wise, it is crucial to note that a pair of parallel arrows into a $B(\!=\!\Rightarrow\!=\!)$-signature $(\Sigma_B, \Sigma_C, \sigma)$ induces equivalence relations on Σ_B and on Σ_C such that σ preserves equivalence.

7.2 The *PART* construction

How can we have a closer look at the difference between the expressiveness of $(=\Rightarrow=)$ and that of $R(R = \Rightarrow R =)$? By Proposition 4.8.1, we can consider $P(\stackrel{f}{=}\Rightarrow\stackrel{f}{=})$ instead of the latter, since both are equivalently expressive.

Kreowski [Kre87] describes a representation of $P(\stackrel{f}{=}\Rightarrow\stackrel{f}{=})$ in $B(=\Rightarrow=)$. His *PART* construction takes a based algebra (B, C, b), where C can be viewed as an extension of B by some "junk", and yields a partial algebra: simply interpret the "junk" values outside $b[B]$ in C as "undefined".

More formally, this is a simple institution representation

$$\mu^{PART} = (\Phi^{PART}, \alpha^{PART}, \beta^{PART}): P(\stackrel{f}{=}\Rightarrow\stackrel{f}{=}) \longrightarrow B(=\Rightarrow=)$$

defined by

- $\Phi^{PART}(S, OP, POP)$ is the based theory $((\Sigma_B, \Sigma_C, \sigma), \Gamma)$. It has a renaming of (S, OP) as base part: $\Sigma_B = (\{s^B \mid s \in S\}, \{op^B : s_1^B, \ldots, s_n^B \longrightarrow s^B \mid op: s_1, \ldots, s_n \longrightarrow s \in OP\})$, while the target part additionally includes the partial operations: $\Sigma_C = (\{s^C \mid s \in S\}, \{op^C : s_1^C, \ldots, s_n^C \longrightarrow s^C \mid op: s_1, \ldots, s_n \longrightarrow s \in OP\dot\cup POP\})$. The signature morphism σ simply maps s^B to s^C and op^B to op^C. Γ consists of $(\forall x, y : s . b_s(x) = b_s(y) \Longrightarrow x = y)$ for each $s \in S$.

- Φ^{PART}, applied to a signature morphism, simply duplicates its components according to the above renaming.

- $\alpha^{PART}_{(S,OP,POP)}(\forall X . e_1 \wedge \ldots \wedge e_n \Longrightarrow e) = (\forall \tilde{X} . \alpha^{PART}[\![e_1]\!] \wedge \ldots \wedge \alpha^{PART}[\![e_n]\!] \Longrightarrow \alpha^{PART}[\![e]\!])$, considered as a Σ_C-sentence, where $\tilde{X} = \{x : s^B \mid x : s \in X\}$
$\alpha^{PART}[\![x : s]\!] = b_s(x)$ (that is, variables are translated to expressions varying over $b[B]$)
$\alpha^{PART}[\![op(t_1, \ldots, t_n)]\!] = op^C(\alpha^{PART}[\![t_1]\!], \ldots, [\![t_n]\!])$
$\alpha^{PART}[\![t_1 = t_2]\!] = [\![\alpha^{PART}[\![t_1]\!] = \alpha^{PART}[\![t_2]\!]]\!]$
$\alpha^{PART}[\![op(t_1, \ldots, t_n) \stackrel{e}{=} t]\!] = [\![op^C(\alpha^{PART}[\![t_1]\!], \ldots, [\![t_n]\!]) = \alpha^{PART}[\![t]\!]]\!]$

- $\beta^{PART}_{(S,OP,POP)}(B,C,b) = ((b_s[B_s])_{s \in S}, (op_C^C)_{op \in OP}, (op_C^C \mid^{b[B]})_{op \in POP})$, where for $op \in POP$, $op_C^C \mid^{b[B]}$ is the (co-)restriction of op_C^C to arguments which yield values in $b[B]$ (it is undefined otherwise). This restriction yields a relative subalgebra in the sense of Section 3.6. Note that since b is a homomorphism, $b[B]$ is closed under the operations op_C^C for $op \in OP$, so here no restriction is needed.

- $\beta^{PART}_\Sigma(h_B, h_C) = h_C \mid^{b[B]}$

7.2. The PART construction

The representation condition can be verified as follows: Valuations $\tilde{\nu}\colon \tilde{X} \longrightarrow (B,C,b)$ are in one-to-one correspondence with valuations $\nu\colon X \longrightarrow \beta_T^{PART}(B,C,b)$, because the variable translations $b_s(x)$ vary over $b_s[B_s]$ and the carriers of $\beta_T^{PART}(B,C,b)$ are the $b_s[B_s]$ as well. The only difference between the $\nu^\#$- and the $\tilde{\nu}^\#$-interpretation of terms is in the treatment of existence equations:

$$\tilde{\nu} \models \alpha^{PART} [\![op(t_1,\ldots,t_n) \stackrel{\mathrm{e}}{=} t]\!] \iff op_C^C(\tilde{\nu}^\#(t_1),\ldots,\tilde{\nu}^\#(t_n)) = \tilde{\nu}^\#(t) \quad (7.1)$$

$$\nu \models op(t_1,\ldots,t_n) \stackrel{\mathrm{e}}{=} t \iff \nu^\#(t_1),\ldots,\nu^\#(t_n) \in \mathrm{dom}\, op_{\beta^{PART}(B,C,b)}$$
$$\text{and } op_{\beta^{PART}(B,C,b)}(\nu^\#(t_1),\ldots,\nu^\#(t_n)) = \nu^\#(t) \quad (7.2)$$

The values of $\nu^\#$ and $\tilde{\nu}^\#$ and of op_C^C and $op_{\beta^{PART}(B,C,b)}$ coincide. A difference may remain only in the domains. Since variables in $\tilde{X}$ vary over $b[B]$ and $b[B]$ is closed under the Σ_C-operations, $\tilde{\nu}^\#(t) \in b[B]$. Therefore, if the right hand-side of (7.1) holds, then $op_C^C(\tilde{\nu}^\#(t_1),\ldots,\tilde{\nu}^\#(t_n)) \in b[B]$ and $\tilde{\nu}^\#(t) \in \mathrm{dom}\, op_{\beta^{PART}(B,C,b)}$, so the right hand-side of (7.2) holds. The converse direction is easy. Thus, the left hand sides are equivalent as well. Extension from atomic to all formulas gives us the representation condition. □

The expressions $b_s(x)$ varying over $b[B]$, that is, over "defined" or non-junk elements only, correspond to "safe" variables in the sense of Antimirov and Degtyarev [AD92b, AD92a]. They are needed for a faithful translation of sentences by μ^{PART}. When considering all based algebras leading to a certain $P(\stackrel{\mathrm{f}}{=}\Rightarrow\stackrel{\mathrm{f}}{=})$-algebra, in some of them junk may satisfy non-trivial equations while in others it may not. Therefore, we cannot relate via the representation condition validity of axioms with unsafe variables (that range over C) in based algebras to validity of axioms in their PART constructions. In fact, there is some "free completion" (see Proposition 7.4.1), where junk is freely generated and thus does not satisfy any non-trivial equation.

So we have to use expressions corresponding to safe variables when translating $P(\stackrel{\mathrm{f}}{=}\Rightarrow\stackrel{\mathrm{f}}{=})$-axioms. Antimirov and Degtyarev examine under which conditions unsafe variables may be used when enrichments shall be consistent. Perhaps this could be used to allow unsafe variables in some of our translations, but note that possibly (properties of) whole theories have to be considered then.

While in $P(\stackrel{\mathrm{f}}{=}\Rightarrow\stackrel{\mathrm{f}}{=})$ there is no possibility to talk positively about the undefined, and all functions are strict, Antimirov and Degtyarev argue that with based theories, one can define non-strict partial functions. Our restriction to based theories generated by $P(\stackrel{\mathrm{f}}{=}\Rightarrow\stackrel{\mathrm{f}}{=})$-theories rules out these non-strict partial functions. On the other hand, proposition 7.7.1 below can be extended to the PART construction on arbitrary based theories. So the power of non-strict partial functions is available in $P(\stackrel{\mathrm{f}}{=}\Rightarrow\stackrel{\mathrm{f}}{=})$, but not in a modular axiom-by-axiom manner. It turns out that non-strictness is handled in $P(\stackrel{f}{=}\Rightarrow\stackrel{f}{=})$ by conditional formulas. This corresponds to the observation that in many formalisms, all non-strict functions are definable by strict functions and non-strict if-then-else.

7.3 Measuring the difference of institutions: categorical retractive representations

If two institutions are (simply) equivalent in expressiveness, they simulate each other with respect to sentences, theories, model categories, theory morphisms, reducts, free constructions etc. in a faithful way such that there is not much space for semantic differences. But clearly, equivalence will be a rare occurrence. For example, $(=\Rightarrow=)$ (resp. $B(=\Rightarrow=)$) and $P(\stackrel{f}{=}\Rightarrow\stackrel{f}{=})$ are not equivalent by Corollary 6.1.5, so partiality cannot be expressed directly in $(=\Rightarrow=)$. But the *PART* construction is a way to view total $B(=\Rightarrow=)$-algebras as partial algebras in $P(\stackrel{f}{=}\Rightarrow\stackrel{f}{=})$.

On the abstract level, we point out that categorical retractive representations may be a reasonable generalization, useful to compare institutions that are not equivalent. They preserve initial models, free constructions and forgetful functors. After introducing the general notion, the *PART* construction is shown to be a categorical retractive representation.

The example of the *PART* construction can be carried over to other cases: Often, constructions β of $\mathcal{I}$-models from $\mathcal{I}'$-models (or "interpretations", "views" of $\mathcal{I}'$-models as $\mathcal{I}$-models) are extensible to institution representations $\mu = (\Phi, \alpha, \beta): \mathcal{I} \longrightarrow \mathcal{I}'$. That is, given an $\mathcal{I}$-theory T, we can find an $\mathcal{I}'$-theory T' that "simulates" T with help of the construction. One example is Equational Type Logic (ETL, [MSS90]), which yields an order-sorted or a partial algebra from an $R(R = \Rightarrow R =)$-model. In [MSS91, p. 97], the authors state that though ETL is a sublogic of $R(R = \Rightarrow R =)$, it has more pragmatic expressiveness gained by interpretation of the binary ETL-predicate as typing. In [SS92], the representation of order-sorted and partial algebras in ETL is formalized as a pre-institution transformation (but an institution representation could be used as well). In the following, we examine some properties of constructions β that are the model part of such institution representations and contain part of the pragmatics mentioned above.

Now, institution representations obtained by such constructions generally fail to be embeddings (otherwise, β is only an isomorphic change of representation, so the construction would have no power). So we have to weaken the notion of embedding while maintaining preservation of as many institution independent language constructs as possible.

7.3. MEASURING THE DIFFERENCE OF INSTITUTIONS: CATEGORICAL RETRACTIVE REPRESENTATIONS

An important thing to be preserved is logical consequence. Arbitrary institution representations do not preserve it, but Cerioli's simulations [Cer93], whose model translation is surjective on objects, do so (Theorem 2.6.1). Often, there is a canonical choice to represent a particular $\mathcal{I}$-model as $\mathcal{I}'$-model plus construction. E.g., a particular $P(\stackrel{f}{=}\Rightarrow\stackrel{f}{=})$-algebra can be represented by a canonical based algebra plus $PART$ construction, see below. Thus, surjectivity of β_T often "comes from" a right inverse β_T^{rinv} with $\beta_T \circ \beta_T^{rinv} = Id$.

Since the following notions only depend on theory categories and model functors, they are developed for specification frames. By applying $\mathbf{Th_0}\colon \mathbf{PlainInst} \longrightarrow \mathbf{SpecFram}$ (or $\mathbf{Th_0}$ starting from $\mathbf{Inst}, \bigwedge \mathbf{Inst}, \mathbf{WeakInst}$), the theory carries over to institutions and plain (simple, conjunctive or weak) representations as well.

Definition 7.3.1 A *retractive specification frame representation* is a specification frame representation $\mu = (\Phi, \beta)\colon \mathcal{F} \longrightarrow \mathcal{F}'$, where β_T has a right inverse up to isomorphism β_T^{rinv} (i. e. $\beta_T \circ \beta_T^{rinv} \cong Id$) for each $\mathcal{F}$-theory T. □

Since retractions are surjective on objects, simple retractive institution representations are a special case of Astesiano's and Cerioli's logical simulations [AC92]. In particular, from the simulations they inherit the nice property of being faithful with respect to logical consequence:

Theorem 7.3.2 Let $\mu\colon \mathcal{I} \longrightarrow \mathcal{I}'$ be a (plain, simple, conjunctive) retractive institution representation between institutions in which model isomorphisms preserve satisfaction. Then semantical entailment is preserved:

$$\Gamma \models_\Sigma \varphi \text{ iff } \alpha_\Sigma[\Gamma] \models'_{\Phi(\Sigma)} \alpha_\Sigma(\varphi)$$

Proof:
Similar to the proof of theorem 2.6.1. □

Considering the three levels of institution equivalence formulated in [Cer93], retractive representations cover the set-theoretical and logical level, but not the categorical level. A categorical representation in the sense of [Cer93] has to preserve initial models. But when using parameterization, module concepts or constraints [ST88a, EM90], preservation not only of initial models, but also of free constructions, forgetful functors and combinations of these is desirable. With categorical retractive representations, this is guaranteed:

Definition 7.3.3 A retractive specification frame representation $\mu = (\Phi, \beta)\colon \mathcal{F} \longrightarrow \mathcal{F}'$ is called *categorical*, if for each $\mathcal{F}$-theory T, the right inverse β_T^{rinv} is also a left adjoint to β_T. □

Proposition 7.3.4 Embeddings are categorical retractive representations. □

Retractive representations can be composed:

Definition 7.3.5 Let $\mu^1 = (\Phi^1, \beta^1, \beta^{1\,rinv}): \mathcal{F} \longrightarrow \mathcal{F}'$ and $\mu^2 = (\Phi^2, \beta^2, \beta^{2\,rinv}): \mathcal{F}' \longrightarrow \mathcal{F}''$ be retractive specification frame representations. Then their composition $\mu = \mu^2 \circ \mu^1: \mathcal{F} \longrightarrow \mathcal{F}''$ consists of the following components:

- $\Phi = \Phi^2 \circ \Phi^1$
- $\beta_T = \beta_T^1 \circ \beta_{\Phi^1 T}^2$
- $\beta_T^{rinv} = \beta_{\Phi^1 T}^{2\,rinv} \circ \beta_T^{1\,rinv}$ □

Proposition 7.3.6 Categorical retractive representations are closed under composition.

Proof: Left adjoints are closed under composition. □

Theorem 7.3.7 Let $\mu = (\Phi, \beta, \beta^{rinv}): \mathcal{F} \longrightarrow \mathcal{F}'$ be a categorical retractive representation. Then μ is compatible with free constructions in the following sense: If $\sigma: T_1 \longrightarrow T_2$ is a theory morphism in $\mathcal{F}$ and $F'^{\Phi\sigma}$ and F^σ exist[1], then

$$F^\sigma \cong \beta_{T_2} \circ F'^{\Phi\sigma} \circ \beta_{T_1}^{rinv}$$

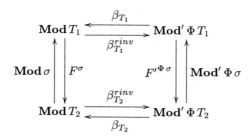

Proof: By naturality of β, we have $\beta_{T_1} \circ \mathbf{Mod}' \Phi\sigma = \mathbf{Mod}\,\sigma \circ \beta_{T_2}$. Turning to left adjoints (which are compositional up to isomorphism), we have

---
[1] The assumption of existence of F^σ can be dropped, see [Mosb].

$F'^{\Phi^\sigma} \circ \beta_{T_1}^{rinv} \cong \beta_{T_2}^{rinv} \circ F^\sigma$. Multiplying left with β_{T_2}, we get $\beta_{T_2} \circ F'^{\Phi^\sigma} \circ \beta_{T_1}^{rinv} \cong F^\sigma$. □

Theorem 7.3.8 If $\mu = (\Phi, \beta, \beta^{rinv}) \colon \mathcal{F} \longrightarrow \mathcal{F}'$ is a categorical retractive representation between liberal specification frames having initial theories (preserved by Φ) with terminal model categories, then initiality of models is preserved by β and β^{rinv}. Note that generally β does not *reflect* initiality.

Proof: β_T^{rinv} is left adjoint and hence preserves colimits, especially initial objects. Considering β_{T_2} for an arbitrary theory T_2, let T_1 and ΦT_1 be initial theories and $\sigma \colon T_1 \longrightarrow T_2$ be the unique theory morphism. Then letting $*$ denote the initial object in each category, we have $\beta_{T_2}(*) = \beta_{T_2}(F'^{\Phi^\sigma}(\beta_{T_1}^{rinv}(*)))$ (since F'^{Φ^σ} and $\beta_{T_1}^{rinv}$, being left adjoints, preserve initiality) $= F^\sigma(*) = *$ (using the diagram above, and again left adjointness). □

Corollary 7.3.9 If $\mu = (\Phi, \beta, \beta^{rinv}) \colon \mathcal{F} \longrightarrow \mathcal{F}'$ is a categorical retractive representation, **Th$_0$** and **Th$'_0$** are cocomplete and **Mod**, **Mod$'$** and Φ preserve colimits, then β and β^{rinv} preserve initial objects. □

7.4 *PART* is categorical retractive

The *PART* construction, more precisely β_T^{PART} for each $P(\stackrel{f}{=}\Rightarrow\stackrel{f}{=})$-theory T, has a left adjoint right inverse, so μ^{PART} is categorical retractive.

For any $P(\stackrel{f}{=}\Rightarrow\stackrel{f}{=})$-theory $T = ((S, OP, POP), \Gamma)$ let $((\Sigma_B, \Sigma_C, \sigma), \Theta) := \Phi^{PART}(S, OP, POP)$. Let $\beta_{((S,OP,POP),\Gamma)}^{PART\text{-}rinv}(A)$ be the based $((\Sigma_B, \Sigma_C, \sigma), \Theta)$-algebra with base part $((A_s)_{s \in S}, (op_A)_{op \in OP})$ and target part $C = ((C_s)_{s \in S}, (op_C)_{op \in \Sigma_C})$, where

- $C := T_{\Sigma_C}(A)/\equiv$ (where each element $a \in A_s$ is used as "variable" $c^a : s$)
- $\equiv_s$ is the Σ_C-congruence generated from $op^C(c^{a_1}, \ldots, c^{a_n}) \equiv_s c^{op_A(a_1, \ldots, a_n)}$ for $a_1, \ldots, a_n \in \text{dom } op_A$ and $op \colon s_1, \ldots, s_n \longrightarrow s \in OP \dot\cup POP$

The homomorphism b between base and target part is defined by $b_s(a) := [c^a]_{\equiv_s}$. $\beta_{((S,OP,POP),\Gamma)}^{PART\text{-}rinv}(A)$ is the "free completion" of A, see [Bur86].

Now if for some $t \in T_{\Sigma_C}(A)_s$, all operation applications in t are defined, $[t]_{\equiv_s} = [c^a]_{\equiv_s}$ for a unique $a \in A_s$ (simply evaluate t using the definition of

≡). Otherwise, $[t]_{\equiv_s} \notin b[A_s]$. Therefore, b is injective, so (A, C, b) satisfies Θ.
Valuations $\nu: X \longrightarrow A$ are in one-to-one correspondence with valuations $\tilde{\nu}: \tilde{X} \longrightarrow \beta_T^{PART\text{-}rinv}(A)$, because the variables in $\tilde{X}$ only vary over $b[A]$, which is isomorphic to A. With an argument similar to that for μ^{PART}, we have that

$$\nu \models op(t_1, \ldots, t_n) \stackrel{e}{=} t \iff \tilde{\nu} \models op(t_1, \ldots, t_n) = t$$

and by extension to all formulas, we have that $\beta_T^{PART\text{-}rinv}(A)$ satisfies $\Phi^{PART}T$, so that $\beta_T^{PART\text{-}rinv}$ is actually a functor

$$\beta_T^{PART\text{-}rinv}: \mathbf{Mod}^{P(\stackrel{f}{=}\Rightarrow\stackrel{f}{=})}T \longrightarrow \mathbf{Mod}^{BCEL}\Phi^{PART}T$$

(This is *not* natural in T, however. Consider the $P(\stackrel{f}{=}\Rightarrow\stackrel{f}{=})$-theories $T1 \stackrel{\sigma}{\longrightarrow} T2$, where $T1$ consists of one sort symbol and $T2$ additionally of one partial operation symbol. Let A be the T_2-algebra with a singleton carrier set and an undefined operation. Then it can be shown that $card((\beta_{T2}^{PART\text{-}rinv}(A))|_{\Phi\sigma}) = \aleph_0$, but $card(\beta_{T1}^{PART\text{-}rinv}(A|_\sigma)) = 1$.)

Proposition 7.4.1 μ^{PART} is categorical retractive, using $\beta^{PART\text{-}rinv}$.

Proof:

1. Retractivity:
   For each $P(\stackrel{f}{=}\Rightarrow\stackrel{f}{=})$-theory T, we have

   $$Id_{\mathbf{Mod}^{P(\stackrel{f}{=}\Rightarrow\stackrel{f}{=})}T} \cong \beta_T^{PART} \circ \beta_T^{PART\text{-}rinv}$$

   The natural isomorphism is given by $\eta_A: A \longrightarrow \beta_T^{PART}\beta_T^{PART\text{-}rinv}(A) = b\,|^{b[A]}$, where $b: A \longrightarrow C$ is the above inclusion of A into its free completion, and $b\,|^{b[A]}$ is the corestriction of b to $b[A] \cong A$, the carrier of the *PART* construction.

2. Categorical retractivity:
   For each $P(\stackrel{f}{=}\Rightarrow\stackrel{f}{=})$-theory T, we have to show that $\beta_T^{PART\text{-}rinv}$ is left adjoint to β_T^{PART}. The unit of the adjunction is η_A. If $h: A \longrightarrow \beta_T^{PART}(B', C', b')$ is a T-homomorphism in $(\Longrightarrow=)$, we have to define a (unique) $\Phi^{PART}T$-homomorphism $h^\#: \beta_T^{PART\text{-}rinv}A \longrightarrow (B', C', b')$ such that

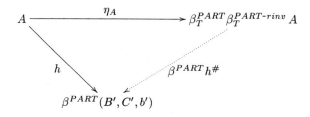

commutes, that is, $\beta_T^{PART}(h^\#) \circ \eta_A = h_C^\# |_{b[A]} \circ b |^{b[A]} = h$. This condition together with the homomorphic property of $h^\#$ forces the following inductive definition of $h_C^\#$:

- $h_C^\#([c^a]) = h_C^\#(b(a)) = h(a)$
- $h_C^\#([op^C(t_1, \ldots, t_n)]) = op_C^C(h_C^\#([t_1]), \ldots, h_C^\#([t_n]))$

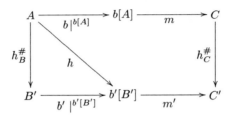

Further, by injectivity of the inclusion m' we must have $b' |^{b'[B']} \circ h_B^\# = h$. Since b' is injective by the Φ^{PART} T-axioms, $b' |^{b'[B']}$ is bijective. This implies that $h_B^\#$ is uniquely determined by h.

Corollary 7.4.2 Logical entailment in $R(R= \Rightarrow R=)$ can be faithfully simulated by logical entailment in $(=\Rightarrow =)$.

Proof: By the proposition, and propositions 4.8.1 and 7.1.1 and theorem 2.6.1.

7.5 Other categorical retractive representations

The representation $\mu^M: R(R= \Rightarrow R=) \longrightarrow UR(R= \Rightarrow R=)$ from Horn Clause Logic to unsorted Horn Clause Logic (see [AC92]) is categorical retractive: the left adjoint right inverse applied to a many-sorted algebra yields the one-sorted algebra, which is freely generated by the many-sorted carriers and ill-sorted operation applications. On the other hand, $R(R= \Rightarrow R=)$ and $UR(R= \Rightarrow R=)$ cannot be equivalent in expressiveness because in $UR(R= \Rightarrow R=)$, all forgetful functors are faithful while in $R(R= \Rightarrow R=)$, this is not the case.

The representation $\mu^D: \mathcal{DNS} \longrightarrow \mathcal{DTL}$ between the institutions of non-strict partial algebras and first-order structures (each with disjunctive sentences) is shown to be categorical retractive in [Cer93, 3.2.48] (where our right inverse is called left inverse, caused by different orders of composition). Note that both institutions are non-liberal.

7.5.1 Level 5 versus level 4

The simple institution representation $\mu^P \colon P(\stackrel{e}{=}\Rightarrow\stackrel{e}{=}) \longrightarrow R(R = \Rightarrow R =)$ which represents partial algebras as total algebras using definedness relations (see [Cer93, 3.2.9]) can be shown to be categorical retractive. Thus it is a link from level 5 to level 4 (see chapter 5). Up to minor differences, this representation also captures the essence of partiality in ETL [MSS90, MSS91] and unified algebras [Mos89], although definedness is coded here in a bit more complex way. This formalizes the situation that ETL-theories and unified algebras are on one hand special Horn Clause Theories, but on the other hand they are interpreted as theories of partial algebras.

A conjunctive categorical retractive institution representation $\mu = (\Phi, \alpha, \beta) \colon P(\stackrel{e}{=}\Rightarrow\stackrel{e}{=}) \longrightarrow ETL$ is defined as follows:

- Φ maps a $P(\stackrel{e}{=}\Rightarrow\stackrel{e}{=})$-signature (S, OP, POP) to the ETL-signature (note that these are unsorted):

    ops op n-ary for $op \colon s_1, \ldots, s_n \longrightarrow s \in OP \cup POP$
    s 0-ary for $s \in S$
    $\forall x_1, \ldots, x_n . \, x_1 \colon s_1 \wedge \cdots \wedge x_n \colon s_n \Longrightarrow op(x_1, \ldots, x_n) \colon s$
    for $op \colon s_1, \ldots, s_n \longrightarrow s \in OP$

- $\beta_{(S,OP,POP)}$ maps a $\Phi(S, OP, POP)$-model M to M' with

    – carriers $M'_s := \{\, a \in |M| \mid a :_M s_M \,\}$ for $s \in S$
    – $op_{M'}(a_1, \ldots, a_n) := op_M(a_1, \ldots, a_n)$ for $op \colon s_1, \ldots, s_n \longrightarrow s \in OP$, $a_i \in M'_{s_i}$ for $i = 1, \ldots, n$. By the axiom in $\Phi(S, OP, POP)$, this is well defined.
    – $pop_{M'}(a_1, \ldots, a_n) :=$
    $\begin{cases} pop_M(a_1, \ldots, a_n), & \text{if } pop_M(a_1, \ldots, a_n) :_M s_M \\ \text{undefined}, & \text{otherwise} \end{cases}$

    while a homomorphism $h \colon M_1 \longrightarrow M_2$ is mapped to $(h \mid_{M'_1,s})_{s \in S}$, which is an S-function from M'_1 to M'_2 since h has to preserve $:$ and the sort constants.

- $\alpha_{(S,OP,POP)}$ maps a (S, OP, POP)-sentence

    $$\forall x_1 \colon s_1, \ldots, x_n \colon s_n . \, t_1 \stackrel{e}{=} u_1 \wedge \cdots \wedge t_n \stackrel{e}{=} u_n \Longrightarrow t \stackrel{e}{=} u$$

    to the set consisting of a $\Phi(S, OP, POP)$-sentence

    $$\forall x_1, \ldots, x_n . \, x_1 \colon s_1 \wedge \cdots \wedge x_n \colon s_n \wedge$$

7.5. OTHER CATEGORICAL RETRACTIVE REPRESENTATIONS

$$\bigwedge_{t' \in Sub} t' : s' \wedge t_1 = u_1 \wedge \cdots \wedge t_n = u_n \implies e$$

for each $e \in \{\, t' : s' \mid t'$ subterm of sort s' of t or $u\,\} \cup \{\, t = u\,\}$, where Sub is the set of all subterms of sort s' of t_1 or ... or t_n or u_1 or ... or u_n.

- To define $\beta^{rinv}_{(S,OP,POP)}$, for a (S, OP, POP)-model M' let N be the free $\Phi(S, OP, POP)$-model over the set $\dot{\bigcup}_{s \in S} M'_s$ with unit $\eta \colon \bigcup_{s \in S} M'_s \longrightarrow |N|$. Define a congruence on N generated by the axioms

$$op_N(\eta(a_1), \ldots, \eta(a_n)) \equiv \eta(op_{M'}(a_1, \ldots, a_n))$$

for $op \colon s_1, \ldots, s_n \longrightarrow s \in OP \cup POP$, $a_i \in M'_{s_i}$ for $i = 1, \ldots, n$ and $op_{M'}(a_1, \ldots, a_n)$ defined. Then $\beta^{rinv}_{(S,OP,POP)}(M')$ is $N/\equiv$.

The representation condition follows from the observation that valuations $\nu \colon \{\, x_1, \ldots, x_n\,\} \longrightarrow M$ satisfying $x_1 : s_1 \wedge \cdots \wedge x_n : s_n$ are in one-one-correspondence with valuations $\nu' \colon \{\, x_1 : s_1, \ldots, x_n : s_n\,\} \longrightarrow \beta_\Sigma(M)$. The proof that β^{rinv} actually is a pointwise left-adjoint right inverse proceeds similar to the proof of Proposition 7.4.1.

7.5.2 Level 4 versus level 3

Let $\mu^{char} = \mu^{flatten} \circ \mu^{PART} \circ \mu^{chardom}$. $\mu^{char} \colon R(R = \Rightarrow R =) \longrightarrow (=\Rightarrow=)$ essentially represents relations by total characteristic functions, where the elements outside a relation are mapped to junk. The determination of what junk is, is achieved by a based algebra construction. Thus we get a link from level 4 to level 3.

7.5.3 Level 3 versus level 2

Conjecture 7.5.1 There is no categorical retractive representation from $(=\Rightarrow=)$ (level 3) to $RP(R \stackrel{f}{=})$ (level 2). Thus there is a gap between conditional and unconditional axioms.

To prove this conjecture, one might try to use the last part of theorem 6.3.2 stating that in $RP(R \stackrel{f}{=})$, specifiable PADTs preserve kernels. Unfortunately, the categorical formulation of this property uses coequalizers and pullbacks, but only one of them is preserved by each member of the adjoint pair translating models in a categorical retractive representation. This does not suffice for proving that preserving kernels is a property being preserved by categorical retractive representations.

7.5.4 Level 2 versus level 1

Let $NMTP(\stackrel{f}{=})$ be the restriction of $P(\stackrel{f}{=})$ to models with non-empty carrier sets. A link from level 2 to level 1 is the following simple retractive representation $\mu^{P=}: NMTP(\stackrel{f}{=}) \longrightarrow (=)$. (By composing it with a restriction of the embedding number 2.4 from $RP(R \stackrel{f}{=})$ to $P(\stackrel{f}{=})$ defined in subsection 5.5, we get a retractive representation from $NMTRP(R \stackrel{f}{=})$ to $(=)$.)

$\mu^{P=}$ maps a signature (S, OP, POP) to the theory consisting of sorts $S \times \{0\} \cup S \times \{1\}$, total operations

$op\colon (s_1, 0) \times \cdots \times (s_n, 0) \longrightarrow (s, 0)$ for $op\colon s_1, \ldots, s_n \longrightarrow s \in OP$
$pop\colon (s_1, 0) \times \cdots \times (s_n, 0) \longrightarrow (s, 1)$ for $pop\colon s_1, \ldots, s_n \longrightarrow s \in POP$
$inj_s\colon (s, 0) \longrightarrow (s, 1)$ for $s \in S$
$proj_s\colon (s, 1) \longrightarrow (s, 0)$ for $s \in S$

For each $s \in S$, one axiom is needed:

$$\forall x\colon (s, 0) .\ proj_s(inj_s(x)) = x$$

Thus for each sort, we add a supersort $(s, 1)$ which shall hold the values for undefined applications.

A $\Phi(\Sigma)$-model M is translated by forgetting all the $M_{(s,1)}$ and taking $M_{(s,0)}$ as carriers $\beta(M)_s$. Total operations can remain the same, while operations pop_M for $pop\colon w \longrightarrow s \in POP$ have to be restricted:

$$pop_{\beta(M)}(a_1, \ldots, a_n) = \begin{cases} proj_M(pop_M(a_1, \ldots, a_n)) \\ \quad \text{if } \exists a \in M_{(s,0)} .\ inj_{s,M}(a) = pop_M(a_1, \ldots, a_n) \\ \text{undefined, otherwise} \end{cases}$$

A sentence $\forall X .\ pop(t_1, \ldots, t_n) \stackrel{f}{=} t$ is mapped to

$$\forall X .\ pop(t_1, \ldots, t_n) = inj_s(t)$$

where s is the sort of t. Note, due to the flatness of partiality in $P(\stackrel{f}{=})$, we end up with a purely total equation.

Now $\beta_\Sigma(M) \models \forall X .\ pop(t_1, \ldots, t_n) \stackrel{f}{=} t$ iff for all valuations $\nu\colon X \longrightarrow \beta_\Sigma(M)$, $pop_{\beta_\Sigma(M)}(\nu^\#(t_1), \ldots, \nu^\#(t_n))$ is defined and $= \nu^\#(t)$ iff for all valuations $\nu\colon X \longrightarrow M$, $pop_M(\nu^\#(t_1), \ldots, \nu^\#(t_n)) = inj_M(\nu^\#(t))$ iff $M \models \alpha_\Sigma(\forall X .\ pop(t_1, \ldots, t_n) \stackrel{f}{=} t)$.

To show retractivity, let $\beta^{rinv}(M)$ be just a copy of M (held in the sorts with 0-label), embedded with the inj-operations into its free completion (held in the sorts with 1-label). $proj$ is the identity on M and takes an arbitrary value outside of M (which is there since the carriers of M are non-empty).

7.6 Specifiability of representations

In the examples of the preceding section, we have an embedding $\mu': \mathcal{F}' \longrightarrow \mathcal{F}$, but only a categorical retractive representation $\mu: \mathcal{F} \longrightarrow \mathcal{F}'$. So we possibly cannot simulate the "loose semantics" of $\mathcal{F}$ already in $\mathcal{F}'$, but by theorem 7.3.7, we can simulate the relevant forgetful and free constructions of $\mathcal{F}$ in $\mathcal{F}'$. So we may say that in some sense $\mathcal{F}$ is not more complex than "$\mathcal{F}'$+construction β". But what about the other way round? Can "$\mathcal{F}'$+construction β" already be expressed in $\mathcal{F}$?

To answer this question, we first define specifiable specification frame representations and then show their closure under composition.

Definition 7.6.1 Let $\mathcal{F} = (\mathbf{Th}, \mathbf{Mod})$ be a specification frame. A functor $F: \mathbf{Mod}\, T_1 \longrightarrow \mathbf{Mod}\, T_2$ is called *specifiable*, if there are theory morphisms $\rho: T_1 \longrightarrow T'$ and $\theta: T_2 \longrightarrow T'$ for some theory T' such that

$$F \cong \mathbf{Mod}\, \theta \circ F^\rho$$

F is called *persistently specifiable*, if F^ρ is persistent, that is, $\mathbf{Mod}\, \rho \circ F^\rho \cong Id_{\mathbf{Mod}\, T_1}$.

Thus a functor is specifiable in this sense if and only if it is expressible as a composite of a free and a forgetful functor. That is, the theory T' may contain "hidden parts", which are forgotten by the forgetful functor $\mathbf{Mod}\, \theta$.

Definition 7.6.2 Let $\mu = (\Phi, \beta): \mathcal{F} \longrightarrow \mathcal{F}'$ be a specification frame representation, and $\mu' = (\Phi', \beta'): \mathcal{F}' \longrightarrow \mathcal{F}$ be an embedding. We call μ *specifiable* w.r.t. μ', if for each theory T in $\mathcal{F}$, there is a theory ΨT "including" T and $\Phi' \Phi T$ (more precisely, there are monomorphisms $\theta T : T \hookrightarrow \Psi T$ and $\rho T : \Phi' \Phi T \hookrightarrow \Psi T$), such that

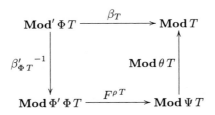

commutes up to natural isomorphism. If in addition μ is a retractive representation, we call μ a *specifiable retractive representation* w.r.t. μ', if μ is specifiable w.r.t. μ' as a specification frame representation and if there is a theory ΞT "including" T and $\Phi' \Phi T$ (that is, there are monomorphisms $\theta T : T \hookrightarrow \Xi T$ and $\rho T : \Phi' \Phi T \hookrightarrow \Xi T$), such that

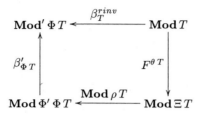

commutes up to natural isomorphism.

Persistent specifiability is defined as expected (i.e. $F^{\rho T}$ resp. $F^{\theta T}$ is required to be persistent). □

If $\mathcal{F}'$ is a subframe of $\mathcal{F}$ and we have a categorical specifiable retractive representation $\mu\colon \mathcal{F} \longrightarrow \mathcal{F}'$, then we have a very precise measure of the difference of the expressive power of $\mathcal{F}$ and $\mathcal{F}'$ with respect to parameterization. The slogan is

$$\text{Power of } \mathcal{F} = \text{Power of } \mathcal{F}' + \text{Power of } \beta + \text{Power of } \beta^{rinv}.$$

That is, the constructions β and β^{rinv} are those parts of $\mathcal{F}$, which already contain the essence of the difference in expressiveness between $\mathcal{F}$ and $\mathcal{F}'$ (with respect to parameterization).

The "$\leq$"-direction of the above "equality" follows from theorem 7.3.7, which allows us to simulate free construction in $\mathcal{F}$ by free constructions in $\mathcal{F}'$ plus β plus β^{rinv}. Considering the "$\geq$"-direction, we have to check that each of the three components of the right hand side has less or equal power than the left hand side. For the first component, this follows from $\mathcal{F}'$ being a subframe of $\mathcal{F}$, and for the other two, it follows from their specifiability in $\mathcal{F}$. The equivalence functor $\beta'_{\Phi T}$ is not taken into account here, because equivalences add no complexity, but only change representation.

Considering compositions of specifiable specification frame representations, we need

Proposition 7.6.3 Let $\mathcal{F} = (\mathbf{Th}, \mathbf{Mod},)$ be a specification frame having amalgamation. Then the extension lemma holds. That is, for each pushout

7.6. Specifiability of representations

in **Th**

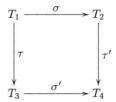

with F^σ persistent, the diagram

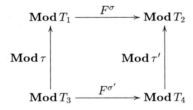

commutes up to natural isomorphism. (This implies that persistently specifiable functors are closed under composition, see below.)

Proof: See [Bau91].

Theorem 7.6.4 Persistently specifiable specification frame representations between specification frames with amalgamation are closed under composition.

Proof: Let $\mu^1: \mathcal{F} \longrightarrow \mathcal{F}'$ and $\mu^2: \mathcal{F}' \longrightarrow \mathcal{F}''$ be two such specification frame representations being specifiable w.r.t. embeddings $\mu^4: \mathcal{F}' \longrightarrow \mathcal{F}$ resp. $\mu^3: \mathcal{F}'' \longrightarrow \mathcal{F}'$, and let $\mu = \mu^2 \circ \mu^1$, $\mu' = \mu^4 \circ \mu^3$. Specifiability of μ^1 resp. μ^2 leads to theory morphisms $\theta^i T: T \longrightarrow \Psi^i T$ and $\rho^i T: \Phi^{5-i} \Phi^i T \longrightarrow \Psi^i T$ for $i = 1, 2$, as in definition 7.6.2.

For an $\mathcal{F}$-theory T, take the following pushout in **Th**

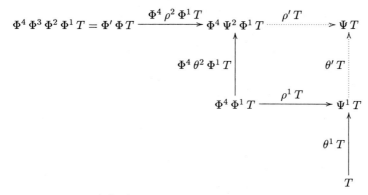

and let $\rho T := \rho' T \circ \Phi^4 \rho^2 \Phi^1 T$ and $\theta T := \theta' T \circ \theta^1 T$.

Commutativity of the outer square in figure 7.6 follows from commutativity of (1) through (9). (1) and (4) commute by definition 7.3.5. Specifiability of μ^1 and μ^2 leads to commutativity (up to isomorphism) of (3) and (2). (5) commutes by naturality of β^4. (6) and (9) commute by functoriality of **Mod** (and compositionality of left adjoints). (7) commutes by theorem 7.3.7, since μ^4 is categorical retractive by 7.3.4, and (8) commutes by proposition 7.6.3. Since the outer square commutes, μ is specifiable via Ψ, θ and ρ.

Theorem 7.6.5 Persistently specifiable retractive representations between specification frames having amalgamation are closed under composition.

Proof: Analogous to theorem 7.6.4. Use figure 7.6 with all arrows reversed, β_T, $\beta^2_{\Phi^1 T}$ and β^1_T replaced by their right inverses, Ψ replaced by Ξ, and F and **Mod** interchanged.

7.7 Specifiability of $PART$

For the $PART$ construction $\mu^{PART} \colon P(\stackrel{f}{=}\Rightarrow\stackrel{f}{=}) \longrightarrow B(=\Rightarrow=)$, we want to show that β^{PART} and $\beta^{PART\text{-}rinv}$ are specifiable. Therefore, we have to interpret $B(=\Rightarrow=)$ as subframe of $P(\stackrel{f}{=}\Rightarrow\stackrel{f}{=})$. By abuse of language, we call the composition of $\mu^{flatten} \colon B(=\Rightarrow=) \longrightarrow (=\Rightarrow=)$ with the obvious inclusion $\mu \colon (=\Rightarrow=) \hookrightarrow P(\stackrel{f}{=}\Rightarrow\stackrel{f}{=})$ also $\mu^{flatten}$.

To show that β^{PART} is specifiable, for each $P(\stackrel{f}{=}\Rightarrow\stackrel{f}{=})$-theory $T = ((S, OP, POP), \Gamma)$, we have to define a $P(\stackrel{f}{=}\Rightarrow\stackrel{f}{=})$-theory $\Psi^{PART} T$ which contains T and $\Phi^{flatten} \Phi^{PART} T$. Let T' be the union of $\Phi^{flatten} \Phi^{PART} T$ with a

7.6. Specifiability of representations

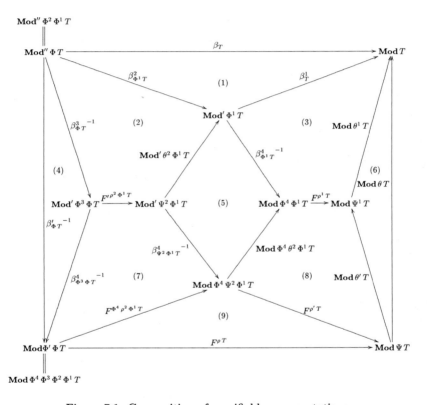

Figure 7.1: Composition of specifiable representations

renaming of T, say $\theta: T \hookrightarrow T'$ and $\rho: \Phi^{flatten} \Phi^{PART} T \subseteq T'$, so we have s^B, s^C and b_s in T' like in the definitions of Φ^{PART} and $\Phi^{flatten}$. Then $\Psi^{PART} T$ is T' augmented by the following total operation symbols

$$make_s : s^B \longrightarrow \theta s \text{ for } s \in S$$

$$toC_s : \theta s \longrightarrow s^C \text{ for } s \in S$$

the partial operation symbols

$$fromC_s : s^C \longrightarrow ? \theta s \text{ for } s \in S$$

and the following axioms:

$$b_s(x) = b_s(y) \implies make_s(x) = make_s(y) \text{ for } s \in S \tag{7.3}$$

$$toC_s(make_s(x)) = b_s(x) \text{ for } s \in S \tag{7.4}$$

$$fromC_s(b_s(x)) \stackrel{e}{=} make_s(x) \text{ for } s \in S \tag{7.5}$$

$$fromC_s(op^C(toC_{s_1}(x_1), \ldots, toC_{s_n}(x_n))) \stackrel{e}{=} \theta\, op(x_1, \ldots, x_n) \tag{7.6}$$

for $op: s_1, \ldots, s_n \longrightarrow s \in OP$

$$pop^C(toC_{s_1}(x_1), \ldots, toC_{s_n}(x_n)) \stackrel{e}{=} b_s(y) \iff \theta\, pop(x_1, \ldots, x_n) \stackrel{e}{=} make_s(y) \tag{7.7}$$

for $pop: s_1, \ldots, s_n \longrightarrow s \in POP$

The intuition behind $\Psi^{PART} T$ is as follows. Since in $P(\stackrel{f}{=} \Rightarrow \stackrel{f}{=})$ we don't have arbitrary existence equations as in $P(\stackrel{e}{=} \Rightarrow \stackrel{e}{=})$, we cannot generate the carriers of $PART(B, C, b)$ with a partial operation from C. So the $make$ operations generate a copy of B, and axioms (7.3) factor this by the kernel of b, so we get the carriers $b[B]$ of $PART(B, C, b)$. Operations toC and $fromC$ are needed for the translation between C and $PART(B, C, b)$, where $fromC$ must be partial, because $PART(B, C, b)$ may have only subsets of C as carriers.

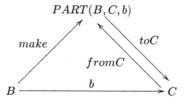

Axioms (7.4) and (7.5) express the relations between these functions, axioms (7.6) translate the operations op^C (for $op \in OP$) to total operations $\theta\, op$ on

7.7. SPECIFIABILITY OF PART

$PART(B,C,b)$ and axioms (7.7) generate the partial restriction of a total Φ^{PART} T-operation pop^C (for $pop \in POP$) to $PART(B,C,b)$.

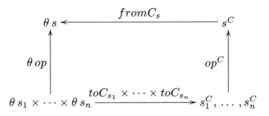

Now we have

Proposition 7.7.1 $\Psi^{PART} T$ actually specifies β^{PART}, that is,

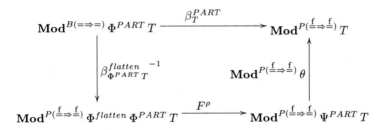

commutes up to natural isomorphism.

Proof: Let (B,C,b) a $\Phi^{PART} T$-algebra, $A' = {\beta^{flatten}_{\Phi^{PART} T}}^{-1}(B,C,b)$ its flattening and $A = \beta^{PART}_T(B,C,b)$ its $PART$-construction. Extend A to a $\Psi^{PART} T$-algebra $\tilde{A}$ as follows:

- $\tilde{A}|_\rho := A'$

- $\tilde{A}|_\theta := A$

- $make_{s,\tilde{A}} := b_{s,A'}|^{A_s} : A'_{s^B} \longrightarrow A_s$

- $toC_{s,\tilde{A}}$ is the inclusion of A_s into A'_{s^C}

- dom $fromC_{s,\tilde{A}} = A_s \subseteq A'_{s^C}$ and $fromC_{s,\tilde{A}}(a) = a$ for $a \in A_s$

Since $make_{s,\tilde{A}} = b_{s,A'}$, axioms (7.3) to (7.7) are checked to hold in $\tilde{A}$ easily.

Let F be the algebra $F^\rho A'$. Since F^ρ is a left adjoint, we have an arrow $id^\#\colon F^\rho A' \longrightarrow \tilde{A}$ adjoint to $id_{A'}\colon A' \longrightarrow \tilde{A}|_\rho$ such that

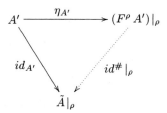

commutes.

Therefore the unit $\eta_{A'}$ is a section (hence injective). Now $\eta_{A'}$ is surjective as well. To show this, we prove by induction on the F-operations (which, together with $\eta_{A'}[A']$, generate F), that the elements of F_{s^B} and F_{s^C} are of form $\eta_{A'}(a)$ for $a \in A'_{s^B}$ resp. $a \in A'_{s^C}$ and the elements of $F_{\theta s}$ are of form

$$make_{s,F} \circ \eta_{A'}(a)$$

for some $a \in A'_{s^B}$. For op^B and op^C, the inductive step is easy: use the homomorphic property of $\eta_{A'}$. For $make_s$, it's trivial. toC_s is handled with (7.4). By (7.5), $fromC_s$ is defined only on values $b_{s,F} \circ \eta_{A'}(a) = \eta_{A'} \circ b_{s,A'}(a)$, which yield the form above.

For $\theta\, op$ ($op \in OP$), we have

$$
\begin{aligned}
&(\theta\, op)_F \circ make_{s_1,\ldots,s_n,F} \circ \eta_{A'}(a) \\
={}& fromC_{s,F} \circ op_F^C \circ toC_{s_1,\ldots,s_n,F} \circ make_{s_1,\ldots,s_n,F} \circ \eta_{A'}(a) \quad \text{by (7.6)} \\
={}& fromC_{s,F} \circ op_F^C \circ b_{s_1,\ldots,s_n,F} \circ \eta_{A'}(a) \quad \text{by (7.4)} \\
={}& fromC_{s,F} \circ b_{s,F} \circ op_F^B \circ \eta_{A'}(a) \quad \text{by } \Phi^{flatten}\text{-axioms} \\
={}& make_{s,F} \circ op_F^B \circ \eta_{A'}(a) \quad \text{by (7.5)}
\end{aligned}
$$

By (7.7), the values of $(\theta\, pop)_F$ (for $pop \in POP$) are in the desired form.

Therefore, $\eta_{A'}$ is an isomorphism with inverse $id^\# |_\rho$. We extend $\eta_{A'}$ to a homomorphism $id^{\#^{-1}}\colon \tilde{A} \longrightarrow F^\rho A'$ (which is the inverse of $id^\#$) by putting

$$id^{\#^{-1}}(make_{s,\tilde{A}}(a)) := make_{s,F} \circ \eta_{A'}(a) \quad (a \in A_s)$$

This is well-defined by the homomorphic property of $\eta_{A'}$. We have to check the homomorphic property of $id^{\#^{-1}}$. The only complicated cases are the operations $\theta\, pop$ for $pop \in POP$: Let $(\theta\, pop)_{\tilde{A}} \circ make_{s,\tilde{A}}(a_1, \ldots, a_n)$ be defined with value $make_{s,\tilde{A}}(a)$ (remember that the $PART$ construction acts on $b_{s,A}[B]$ and $b_{s,A'} = make_{s,\tilde{A}}$ as a function). By the construction of β^{PART}, this means that $pop_{\tilde{A}}^C \circ b_{s_1,\ldots,s_n,A'}(a_1, \ldots, a_n)$ is defined with value $b_{s,A'}(A)$. Since $\eta_{A'}$ is a

7.7. SPECIFIABILITY OF PART

$\Phi^{flatten} \Phi^{PART}$ T-homomorphism, this means $pop_F^C \circ b_{s_1,\ldots,s_n,F} \circ \eta_{A'}(a_1,\ldots,a_n)$ is defined with value $b_{s,F} \circ \eta_{A'}(a)$. By (7.4), we have $pop_F^C \circ toC_{s_1,\ldots,s_n,F} \circ make_{s_1,\ldots,s_n,F} \circ \eta_{A'}(a_1,\ldots,a_n) = b_{s,F} \circ \eta_{A'}(a)$. By (7.7), this implies

$$(\theta\, pop)_F \circ make_{s_1,\ldots,s_n,F} \circ \eta_{A'}(a_1,\ldots,a_n) = make_{s,F} \circ \eta_{A'}(a)$$

So we have

$$(\theta\, pop)_F \circ id^{\#^{-1}} \circ make_{s_1,\ldots,s_n,\tilde{A}}(a_1,\ldots,a_n) = id^{\#^{-1}} make_{s,\tilde{A}}(a)$$

This shows that $\tilde{A}$ and F are isomorphic, and so must be $\tilde{A}\,|_\theta = A = \beta_T^{PART}(B,C,b)$ and $F|_\theta = (F^\rho\, A')|_\theta = \mathbf{Mod}\,\theta \circ F^\rho \circ \beta_{\Phi^{PART}\,T}^{flatten\ \ -1}(B,C,b)$.

To show that $\beta^{PART\text{-}rinv}$ is specifiable as well, for each $P(\stackrel{f}{=}\Rightarrow\stackrel{f}{=})$-theory $T = ((S, OP, POP), \Gamma)$, we have to define a $P(\stackrel{f}{=}\Rightarrow\stackrel{f}{=})$-theory $\Xi^{PART}\, T$ which contains T and $\Phi^{flatten}\Phi^{PART}\, T$. Let T' be the union of $\Phi^{flatten}\Phi^{PART}\, T$ with a renaming of T, say $\theta: T \hookrightarrow T'$ and $\Phi^{flatten}\Phi^{PART}\, T \subseteq T'$, so we have s^B, s^C and b_s in T' like in the definitions of Φ^{PART} and $\Phi^{flatten}$. Then $\Xi^{PART}\, T$ is T' augmented by the following total operation symbols

$$make_s: \theta\, s \longrightarrow s^B \text{ for } s \in S$$

and the following axioms:

$$op^B(make_{s_1,\ldots,s_n}(x_1,\ldots,x_n)) = make_s(\theta\, op(x_1,\ldots,x_n)) \qquad (7.8)$$

$$\text{for } op: s_1,\ldots,s_n \longrightarrow s \in OP$$

$$op^C(b_{s_1,\ldots,s_n}(make_{s_1,\ldots,s_n}(x_1,\ldots,x_n))) = b_s(make_s(\theta\, op(x_1,\ldots,x_n))) \qquad (7.9)$$

$$\text{for } op: s_1,\ldots,s_n \longrightarrow s \in OP$$

$\theta\, pop(x_1,\ldots,x_n) \stackrel{e}{=} y \Longrightarrow$

$$pop^C(b_{s_1}(make_{s_1}(x_1)),\ldots,b_{s_1}(make_{s_1}(x_1))) = b_s(make_s(y)) \qquad (7.10)$$

$$\text{for } pop: s_1,\ldots,s_n \longrightarrow s \in POP$$

The proof that this actually specifies $\beta^{PART\text{-}rinv}$ is left to the reader. □

The retractive representation $\mu^P: P(\stackrel{e}{=}\Rightarrow\stackrel{e}{=}) \longrightarrow R(R=\Rightarrow R=)$ from section 7.5.1 is also specifiable. We conjecture that $\mu^D: \mathcal{DNS} \longrightarrow \mathcal{DTL}$ from section 7.5 is specifiable as well, whereas $\mu^M: R(R=\Rightarrow R=) \longrightarrow UR(R=\Rightarrow R=)$ (from the same section) does not seem to be specifiable, because the division of one (unsorted) carrier in several (many-sorted) carriers cannot be specified in $R(R=\Rightarrow R=)$.

To show specifiability of $\mu^{chardom}$, μ^{graph} from Section 4.8 and of μ^{embed} and $\mu^{flatten}$ (as retractive representations), the theory $\Psi T = \Xi T$ is built from a renaming $\theta T: T \longrightarrow T'$ and $\Phi' \Phi T$ by adding total operations

$$to_s: s \longrightarrow \theta\, s$$

$$from_s: \theta\, s \longrightarrow s$$

with the axioms
$$to_s(from_s(x)) = x$$

$$from_s(to_s(x)) = x$$

$$\theta\, op(x_1, \ldots, x_n) = to_s(op(from_{s_1}(x_1), \ldots, from_{s_n}(x_n)))$$
for $op: s_1, \ldots, s_n \longrightarrow s \in OP$

specifying to and $from$ to be isomorphisms and the following specific axioms:

$\mu^{chardom}$	$\theta\, R(x_1, \ldots, x_n) \iff$ $G^{\chi_R}(from_{s_1}(x_1),$ $\ldots, from_{s_n}(x_n), from_{s_1}(x_1))$	$R: s_1, \ldots, s_n \in REL$
μ^{graph}	$\theta\, pop(x_1, \ldots, x_n) \stackrel{e}{=} y \iff$ $\chi_{G^{pop}}(from_{s_1}(x_1),$ $\ldots, from_{s_n}(x_n),\ from_s(y)) \stackrel{e}{=}$ $from_{s_1}(x_1)$	$pop: s_1, \ldots, s_n \longrightarrow s \in POP$
μ^{embed}	none	
$\mu^{flatten}$	$\theta\, b_s(to_s(x)) = to_s(b_s(x))$	$s \in \Sigma_B$

By proposition 7.3.6 and theorem 7.6.5, $\mu^{char} = \mu^{flatten} \circ \mu^{PART} \circ \mu^{chardom}$ and $\mu^{char} \circ \mu^P$ are specifiable categorical retractive representations, too.

7.8 Summary

We have examined the expressiveness of the generation of partial algebras from total based algebras with the *PART* construction. Following the remarks after definition 7.6.2, this turns out to be equivalent in expressiveness (with respect to the power of parameterization and module concepts) to Horn Clause Logic. The based-algebra approach with the *PART* construction has a clear intuition behind the formalism while partiality via $R(R= \Rightarrow R=)$ can benefit from the well-developed theory about Horn Clause Theories, see [Pad88]. We hope that, by the equivalence shown, the two approaches can have a fruitful influence on each other. See [Pad88, p. 22f.] for an instructive example.

When considering loose semantics, there is a gap between Conditional Equational Logic and Horn Clause Logic. Another view of the *PART* construction is that it bridges this gap, being a representation of Horn Clause Logic within

7.8. Summary

Conditional Equational Logic, logically faithful and compatible with initial semantics and free constructions.

Altogether, we can state the "equation"

> Horn Clause Logic =
> Based Conditional Equational Logic
> + *PART* construction with inverse

This situation can be described and generalized to other institutions using the notion of (specifiable) categorical retractive representation. In particular, partiality in ET logic and unified algebras can be viewed as categorical retractive representations.

Considering the five levels of expressiveness introduced in chapters 5 and 6, there are (categorical) retractive representations from level 5 to level 4, from level 4 to level 3 and from level 2 to level 1. This shows that level 3, 4 and 5 on one hand and level 1 and 2 on the other hand are equivalent[2] w. r. t. to entailment: we can re-use theorem proving methods like conditional term rewriting, which work for conditional equational logic (level 3), also for level 4 and 5, and we re-use equational term rewriting, which works for level 1, also for level 2.

Between institutions with conditional axioms (level 3 to 5) and institutions with unconditional axioms only (level 1 and 2) there seems to be a gap that cannot be bridged by (categorical) retractive representations either.

---

[2] Well, modulo the fact that some embeddings within the levels are only weak.

Chapter 8

Parchments — a device for combining logics

> "The motivation for developing systems which are not dependent on any particular logical system is that no single logical system is clearly adequate for all purposes. In most practical situations it is necessary to use heterogeneous logics which contain constructs for dealing with different features of computation — parallelism, side-effects, non-determinism, polymorphism, higher-order functions, etc. Some of these features are still not adequately understood on the algebraic level, even when considered in isolation from other features. It makes sense to develop complex logics from simpler logics in a structured fashion, just as complex specifications are built in a structured fashion from simpler units. Methods for building complex logics in this fashion are only beginning to be explored." *D. Sannella and A. Tarlecki* [ST93]

The results of the previous chapters are by no means complete. First, there are some gaps left in the study of embeddability within the five levels of institutions I have introduced, and, more importantly, other institutions have not been considered at all. One would like to know what happens if the concepts that are already there (partiality, relations, order-sortedness, sort constraints, conditional existence-equations) are combined in other possible variations, and, of course, if entirely new concepts are added.

I have tried to design a systematical notation for institutions used throughout the previous chapters. But it has no semantical counterpart: When constructing an institution of partial algebras, one usually redefines every concept for total algebras from scratch, and generalizes it to the partial case. This may become a tedious process. Moreover, one often has to choose between different generalizations, but the design decisions are scattered all over the definitions.

In this chapter, I try to systematize the combination of different institutions. Goguen and Burstall have argued [GB92, BG77] that combinations can be done via (co-)limits. Now the category of institutions and institution morphisms [GB92] is complete [Tar86a], but limits in this category only put together different concepts *without* true interaction.

Therefore, I move to Goguen's and Burstall's parchments [GB85], which allow defining institutions using

- initial algebra semantics for the definition of sentence syntax,
- a special signature for the definition of models and
- a semantical algebra of sentence evaluators for the definition of semantical evaluation of sentences.

I introduce a notion of morphism between parchments, which allows us to combine different concepts *with* interaction, using limits.

Now there is the problem that, in many cases, the limit contains unwanted "junk" truth values and thus the space of truth values need not be preserved. This problem can be solved by imposing a congruence relation on the semantical algebra of the limit. The congruence can usually be generated from a small number of axioms, which state how different concepts interact.

Thus, we cannot expect a combination to be done automatically (as in the case of combining theories), but we have to use a two-step semi-automatic method: first specify the components to be combined, and then specify the interaction of concepts.

These ideas are illustrated by generalizing several concepts from total to partial algebras. The role of the congruence relation here is to decide whether we want a two- or three-valued logic and to determine which concept of equality to use.

This chapter was published in [Mos96c].

8.1 Why institution morphisms do not suffice

As argued in section 1.5, there are different notions of morphisms between institutions (see [Cer93] for an overview). For the combination of logical systems, representations are not adequate. As [Tar96a] points out, institution morphisms from [GB92] are better-suited here, since they capture the intuition of building

8.1. Why institution morphisms do not suffice

a more complex institution I over a simpler institution J. The idea is that J may contain some "basic" concepts, and I may add some further concepts to J. An *institution morphism* $\mu = (\Phi, \alpha, \beta): I \longrightarrow J$ consists of

- a functor $\Phi: \mathbf{Sign}^I \longrightarrow \mathbf{Sign}^J$,
- a natural transformation $\alpha: sen^J \circ \Phi \longrightarrow sen^I$ and
- a natural transformation $\beta: \mathbf{Mod}^I \longrightarrow \mathbf{Mod}^J \circ \Phi^{op}$

such that the following satisfaction invariant holds:

$$\beta_\Sigma(M') \models^J_{\Phi\Sigma} \varphi \iff M' \models^I_\Sigma \alpha_\Sigma(\varphi).$$

It is easy to define a composition and identities, such that we get a category of institutions and institution morphisms. We then have:

Theorem 8.1.1 (Tarlecki[Tar86a]) The category **Ins** of institutions and institution morphisms is complete. □

Thus we can apply the idea of combining things via colimits to institutions themselves, with the special point that we have to take limits here instead of colimits. Taking limits in **CAT** results in categories of "amalgamated objects", i. e. we put signatures and models together at the level of single objects. In contrast to this, sentences are combined with colimits in **Set** (due to the contravariant direction of the sentence component). That is, *sets of* sentences are combined. To show how this works, we introduce some well-known institutions and morphisms between them.

Example 8.1.2 The institution () of many-sorted algebras without sentences is obtained as a subinstitution of (=). □

Example 8.1.3 The institution (=) introduced in section 3.2. □

Example 8.1.4 The institution $P()$ of partial many-sorted algebras without sentences is obtained as a subinstitution of $P(\stackrel{e}{=})$. □

Example 8.1.5 The institution morphism $\mu^= = (Id, \alpha^=, Id): (=) \longrightarrow ()$ is the identity on signatures and models, and for sentences, $\alpha^=_\Sigma$ is the inclusion. □

Example 8.1.6 The institution morphism $\mu^P = (\Phi^P, Id, \beta^P): P() \longrightarrow ()$ maps a signature $\Sigma = (S, OP, POP)$ to $\Phi^P(\Sigma) = (S, OP)$. Now a Σ-model consists of a $\Phi^P(\Sigma)$-model plus some family of partial operations, and β^P_Σ just forgets this family of partial operations. □

We are now prepared to consider the combination of $(=)$ and $P()$ via the pullback

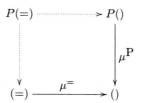

It turns out that signatures and models in $P(=)$ are those of $P()$, while (S, OP, POP)-sentences are (S, OP)-sentences in $(=)$. Thus all equations in $P(=)$ are built up from total operation symbols, while a true interaction of concepts should allow us to write down equations containing partial operation symbols as well, as we are used to from $P(\stackrel{e}{=})$.

8.2 Parchments

The lack of interaction of concepts when taking limits of institutions shows that it is not enough to have a colimit (for example, a disjoint union) of sentence sets. We rather would like to unite *constructions*, or *operators* on sentence sets.

The idea behind the notion of parchment introduced by Goguen and Burstall [GB85] is to generate sentences by initial algebra semantics. For each signature Σ, an abstract syntax $Lang\,\Sigma$, given by some sorts and operations, then generates the Σ-sentences which are considered as terms over $Lang\,\Sigma$.

A combination of abstract syntaxes of sentences then causes a true interaction of different syntactical operators in the generation process. This is also reflected in the semantics. The semantics is given by

- a special signature Γ acting as a semantical universe, such that models are signature morphisms into Γ and
- a $Lang\,\Gamma$-algebra $\mathcal{G}$ specifying semantical evaluation.

Let $\mathbf{Sign}^{R()^*}$ be the category of many-sorted relational signatures from $R(R=)$

8.2. Parchments

with a distinguished sort $*$ and a distinguished relation symbol $D : *$ and signature morphisms preserving $*$ and D.

A *parchment* $P = (\mathbf{Sign}^P, Lang^P, \Gamma^P, \mathcal{G}^P)$ consists of

- a category $\mathbf{Sign}^P$ of signatures,
- a functor $Lang^P \colon \mathbf{Sign}^P \longrightarrow \mathbf{Sign}^{R()*}$,
- a signature Γ^P in $\mathbf{Sign}^P$ and
- a semantical algebra $\mathcal{G}^P$ in $\mathbf{Mod}^{R()*} Lang^P \Gamma^P$.

$\mathcal{G}^P_*$ is the *space of truth values* of P. Usually, $\mathcal{G}^P_* = \mathbf{Bool}$ (with $\mathbf{Bool} = \{\textit{true}; \textit{false}\}$) is required, but for being able to construct limits later on, and also for capturing multi-valued logics, we drop this restriction. But since logical consequence is based on an ultimatively two-valued notion of satisfaction, $D_{\mathcal{G}^P} \subseteq \mathcal{G}^P_*$ singles out the *designated truth values* in P. For example, Kleene's three-valued logic [Kle52] and paraconsistent three-valued logic [Pri79] only differ in the set of designated truth values.

A parchment $P = (\mathbf{Sign}^P, Lang^P, \Gamma^P, \mathcal{G}^P)$ yields an institution $I(P) = (\mathbf{Sign}^P, sen, \mathbf{Mod}, \models)$, where

- $sen\, \Sigma = (T_{Lang^P \Sigma})_*$, where $T_{Lang^P \Sigma}$ is the initial $Lang^P \Sigma$-algebra,
- $sen\,(\sigma\colon \Sigma \longrightarrow \Sigma') = (T_{Lang^P \sigma})_*$,
  where $T_{Lang^P \sigma}\colon T_{Lang^P \Sigma} \longrightarrow T_{Lang^P \Sigma'} \mid_{Lang^P \sigma}$ is the initial homomorphism,
- $\mathbf{Mod}\, \Sigma = \{\, m\colon \Sigma \longrightarrow \Gamma \in \mathbf{Sign}^P(\Sigma, \Gamma)\,\}$ [1],
- $\mathbf{Mod}\, \sigma(\Sigma' \xrightarrow{m} \Gamma) = \Sigma \xrightarrow{\sigma} \Sigma' \xrightarrow{m} \Gamma$, and
- $m \models_\Sigma \varphi$ iff $m^\natural_*(\varphi) \in D_{\mathcal{G}^P}$,
  where $m^\natural \colon T_{Lang^P \Sigma} \longrightarrow \mathcal{G}\mid_{Lang^P m}$ is the initial homomorphism.

With this definition, $\mathbf{Mod}$ always becomes continuous, but there is a slightly different formulation (dropping Γ) which also models non-continuous $\mathbf{Mod}$.

Example 8.2.1 The institution of many-sorted algebras can be generated by a parchment $(\mathbf{Sign}, Lang, \Gamma, \mathcal{G})$, also denoted by ().

- $\mathbf{Sign}$ is taken from the institution (),
- Γ has as sorts all sets $s \in |\mathbf{Set}|$, and as operation symbols $op\colon s_1, \ldots, s_n \longrightarrow s$ all functions $op\colon s_1 \times \cdots \times s_n \longrightarrow s \in |\mathbf{Set}|$ [2],

---

[1] These classes are taken as discrete model categories. Model morphisms can be modeled by 2-cells when $\mathbf{Sign}^P$ is assumed to be a 2-category.

[2] Foundational problems with this construction are discussed in section A.5.

- **Lang** takes a ()-signature (S, OP) to the (overloaded[3]) ()∗-signature
  sorts ∗
  $Term(X, s)$ for X an S-sorted variable system, $s \in S$
  ops $true : *$
  $x : Term(X, s)$ for $x : s \in X$
  $op : Term(X, s_1) \times \cdots \times Term(X, s_n) \longrightarrow Term(X, s)$
  for X S-sorted, $op : s_1, \ldots, s_n \longrightarrow s \in OP$,

- **Lang** takes a signature morphism $\sigma: \Sigma \longrightarrow \Sigma'$ to the signature morphism $Lang(\sigma): Lang(\Sigma) \longrightarrow Lang(\Sigma')$ which maps ∗ to ∗, $Term(X, s)$ to $Term(\sigma(X), \sigma(s))$ where $\sigma(X)$ is defined as in section 3.1, D to D, x to x and op to $\sigma(op)$,

- $\mathcal{G}_* = \mathbf{Bool}$, $D_{\mathcal{G}} = \{true\}$,

- $\mathcal{G}_{Term(X,s)} = [Env(X) \longrightarrow s]$
  where $Env(X)$ is the set of all valuations v of X in S with $v(x : s) \in s$,

- $x_{\mathcal{G}} = \lambda v.v(x)$ for $x : s \in X$, and

- $op_{\mathcal{G}}(f_1, \ldots, f_n) = \lambda v.op(f_1(v), \ldots, f_n(v))$ for $op: s_1, \ldots, s_n \longrightarrow s \in \Gamma$.
  □

Example 8.2.2 (=) is $(\mathbf{Sign}, Lang, \Gamma, \mathcal{G})$ with

- $\mathbf{Sign} = \mathbf{Sign}^{()}$, $\Gamma = \Gamma^{()}$,

- $Lang(S, OP) = Lang^{()}(S, OP)$ **then**
  sorts $Formula(X)$ for X an S-sorted variable system
  ops $=: Term(X, s)^2 \longrightarrow Formula(X)$ for X S-sorted, $s \in S$
  $bind: Formula(X) \longrightarrow *$ for X S-sorted,

- $\mathcal{G}|_{Lang^{()} \Gamma} = \mathcal{G}^{()}$,

- $\mathcal{G}_{Formula(X)} = [Env(X) \longrightarrow \mathbf{Bool}]$,

- $=_{\mathcal{G}} (f_1, f_2) = \lambda v. \begin{cases} true, & \text{if } f_1(v) = f_2(v) \\ false, & \text{if } f_1(v) \neq f_2(v) \end{cases}$, and

- $bind_{\mathcal{G}}(f) = \begin{cases} true, & \text{if } \forall v \in Env(X) : f(v) = true \\ false, & \text{if } \exists v \in Env(X) : f(v) = false \end{cases}$ □

---

[3] Note that $Lang(\Sigma)$ may be overloaded due to overloading in Σ and further because each variable system leads to a new operation with same name.

8.3. Parchment morphisms

Example 8.2.3 The partial many sorted parchment $P() = (\mathbf{Sign}, Lang, \Gamma, \mathcal{G})$:

- **Sign** is taken from the institution $P()$,
- Γ has as sorts all sets, and as total (partial) operation symbols all total (partial) functions between the corresponding sets,
- $Lang(S, OP, POP) = Lang^{()}(S, OP)$ then
  ops $pop : Term(X, s_1) \times \cdots \times Term(X, s_n) \longrightarrow Term(X, s)$
  for X S-sorted, $pop\colon s_1, \ldots, s_n \longrightarrow s \in POP$,
- $\mathcal{G}_* = \mathbf{Bool}$, $D\mathcal{G} = \{\,true\,\}$,
- $\mathcal{G}_{Term(X,s)} = [Env(X) \rightharpoonup s]$ (the set of all partial maps from $Env(X)$ to s),
- $x_{\mathcal{G}}(v) = \begin{cases} v(x), & \text{if } v(x) \text{ is defined} \\ \text{undefined}, & \text{otherwise,} \end{cases}$
- $op_{\mathcal{G}}(f_1, \ldots, f_n)(v) = \begin{cases} op(f_1(v), \ldots, f_n(v)), & \text{if } f_1(v), \ldots, f_n(v) \text{ def.} \\ \text{undefined}, & \text{otherwise,} \end{cases}$
- $pop_{\mathcal{G}}(f_1, \ldots, f_n)(v)$
$= \begin{cases} pop(f_1(v), \ldots, f_n(v)), & \text{if } f_1(v), \ldots, f_n(v) \text{ def. and} \\ & pop(f_1(v), \ldots, f_n(v)) \text{ def.} \\ \text{undefined}, & \text{otherwise.} \end{cases}$

$\square$

We can justify the overloading of names by the fact that I maps $()$ to $()$, $(=)$ to $(=)$ and $P()$ to $P()$.

8.3 Parchment morphisms

The intuition when setting up a parchment morphism $(\Phi, \alpha, g)\colon P \longrightarrow P'$ is that P is more expressive than P', or built over P', adding some new concepts. Φ forgets the new parts of signatures, while α injects the old abstract syntax of sentences into the new one, and g injects the old semantics into the new one.

Definition 8.3.1 Given two parchments $P = (\mathbf{Sign}, Lang, \Gamma, \mathcal{G})$ and $P' = (\mathbf{Sign}', Lang', \Gamma', \mathcal{G}')$, a *parchment morphism*[4] $(\Phi, \alpha, g)\colon P \longrightarrow P'$ consists of

- a functor $\Phi\colon \mathbf{Sign} \longrightarrow \mathbf{Sign}'$ such that $\Phi\,\Gamma = \Gamma'$,

---
[4] Andrzej Tarlecki [ST] has a different notion of parchment morphism.

- a natural transformation $\alpha: Lang' \circ \Phi \longrightarrow Lang$, and
- a $Lang'\, \Gamma'$-homomorphism $g: \mathcal{G}' \longrightarrow \mathcal{G}|_{\alpha_\Gamma}$.

This gives a category **Parch** of parchments and parchment morphisms.
If $g_* = Id$ and $D_\mathcal{G} = D_{\mathcal{G}'}$, then (Φ, α, g) *preserves truth*. (We always have that $g_*[D_{\mathcal{G}'}] \subseteq D_\mathcal{G}$, but non-designation of truth values need not be preserved.) Let **TruthParch** be the category of parchments and parchment morphisms preserving truth. □

Proposition 8.3.2 *I can be extended to a functor $I:$ **TruthParch** $\longrightarrow$ **Ins**.*

Proof:
Put $I((\mathbf{Sign}, Lang, \Gamma, \mathcal{G}) \xrightarrow{(\Phi, \alpha, g)} (\mathbf{Sign}', Lang', \Gamma', \mathcal{G}')) =$

$$I(\mathbf{Sign}, Lang, \Gamma, \mathcal{G}) \xrightarrow{(\Phi, \bar{\alpha}, \beta)} I(\mathbf{Sign}', Lang', \Gamma', \mathcal{G}')$$

with

- $\bar{\alpha}_\Sigma = init_*$, where $T_{Lang'\, \Phi\Sigma} \xrightarrow{init} T_{Lang\, \Sigma}|_{\alpha_\Sigma}$ is the initial homomorphism, and
- $\beta_\Sigma(\Sigma \xrightarrow{m} \Gamma) = \Phi\Sigma \xrightarrow{\Phi m} \Phi\Gamma \xrightarrow{id} \Gamma'$.

Now $\beta_\Sigma(m) \models_{\Phi\Sigma} \varphi'$ iff $(\Phi m)^\natural_*(\varphi') \in D_{\mathcal{G}'}$ iff $g_*((\Phi m)^\natural_*(\varphi')) \in D_\mathcal{G}$ iff $(m)^\natural_*(init_*(\varphi')) \in D_\mathcal{G}$ (by the diagram below) iff $m \models_\Sigma \bar{\alpha}_\Sigma\, \varphi'$.

$$\begin{array}{ccc}
T_{Lang'\,\Phi\Sigma} & \xrightarrow{(\Phi m)^\natural} & \mathcal{G}'|_{Lang'\,\Phi m} \\
{\scriptstyle init}\downarrow & & \downarrow{\scriptstyle g|_{Lang'\,\Phi m}} \\
T_{Lang\,\Sigma}|_{\alpha_\Sigma} & \xrightarrow[(m)^\natural|_{\alpha_\Sigma}]{} \mathcal{G}|_{Lang\, m \circ \alpha_\Sigma} \xrightarrow[Id]{} & \mathcal{G}|_{\alpha_\Gamma \circ Lang'\,\Phi m}
\end{array}$$

The diagram commutes because $T_{Lang'\,\Phi\Sigma}$ is initial. □

Now the institution morphisms $\mu^=$ and μ^P can be lifted to parchment morphisms (preserving truth):

Example 8.3.3 Let $\mu^=: (=) \longrightarrow ()$ be (Id, α, Id), where α_Σ is the inclusion. □

Example 8.3.4 Let $\mu^{\mathbf{P}}\colon P() \longrightarrow ()$ be (Φ, α, g) with

- $\Phi(S, OP, POP) = (S, OP)$,
- α_Σ is the inclusion, and
- $g_{Term(X,s)}$ is the inclusion of $[Env(X) \longrightarrow s]$ into $[Env(X) \rightharpoonup s]$, and g is the identity for the remaining sorts. □

To combine $(=)$ and $P()$, we would like to take a pullback now. But first, we have to prove completeness.

8.4 Putting parchments together using limits

When combining parchments, a natural condition to require would be that the space of truth values is preserved. That is, we want to take limits in **TruthParch**. But **TruthParch** is *not* complete (see Proposition 8.4.8 below)! Therefore, we have to take limits in **Parch**, at the risk of getting parchment morphisms *not* preserving truth, even when starting with a diagram in **TruthParch**.

We now need some prerequisites. The following theorem is Theorem 2 in [TBG91]:

Theorem 8.4.1 If $C\colon Ind^{op} \longrightarrow \mathbf{CAT}$ is an indexed category such that

1. Ind is cocomplete,
2. for each $i \in |Ind|$, $C(i)$ is cocomplete, and
3. for each $\sigma\colon i \longrightarrow j \in Ind$, $C(\sigma)\colon C(j) \longrightarrow C(i)$ has a left adjoint,

then $Flat(C)$ is cocomplete.

To be able to apply Theorem 8.4.1, we need four Lemmas: First

Lemma 8.4.2 Let $\mathbf{Mod}^{R()*}\colon \mathbf{Sign}^{R()*} \longrightarrow \mathbf{CAT}$ be the model functor in $R(R=)$, and let $\Sigma \in \mathbf{Sign}^{R()*}$. Then $\mathbf{Mod}^{R()*}(\Sigma)$ is cocomplete. □

Proof:
$\mathbf{Mod}^{R()*}(\Sigma)$ is locally finitely presentable by Corollary 5.1.6 and thus cocomplete by Definition A.3.2. □

then

Lemma 8.4.3 Let σ be a signature morphism in $\mathbf{Sign}^{R()*}$. Then $\mathbf{Mod}^{R()*}(\sigma)$ has a left adjoint, denoted by F_σ. □

Proof:
Follows from corollary 5.1.4. □

then Lemma 4.1 of [Tar86a]:

Lemma 8.4.4 Let $\mathbf{T}$ be a category and let $INTO(\mathbf{T})$ be the category of functors $F: \mathbf{I} \longrightarrow \mathbf{T}$ into $\mathbf{T}$ with morphisms $(\Phi, \alpha): (I \xrightarrow{F} \mathbf{T}) \longrightarrow (I' \xrightarrow{F'} \mathbf{T})$ consisting of $\Phi: \mathbf{I} \longrightarrow \mathbf{I'}$ and $\alpha: F' \circ \Phi \longrightarrow F$. Then $INTO(\mathbf{T})$ is complete whenever $\mathbf{T}$ is cocomplete. □

and finally Lemma 4.2 of [Tar86a]:

Lemma 8.4.5 Let $\underline{A}$, $\underline{B}$ and $\underline{C}$ be categories and $F: \underline{A} \longrightarrow \underline{C}$, $G: \underline{B} \longrightarrow \underline{C}$ be functors. If $\underline{A}$ and $\underline{B}$ are cocomplete and F is cocontinuous, then the comma category $(F \downarrow G)$ is cocomplete. □

With this, we can prove

Lemma 8.4.6 Let $Proj: INTO(\mathbf{Sign}^{R()*}) \longrightarrow \mathbf{CAT}$ map a functor to its domain category. Let $\mathbf{1}$ be the category with one object and one morphism. Then the comma category $(\mathbf{1} \downarrow Proj)$ is complete.

Proof:
$\mathbf{Sign}^{()}$ is shown to be cocomplete in [TBG91]. Let Σ_* be the signature consisting of $true : *$. Then $\mathbf{Sign}^{R()*}$ is the comma category $(\Sigma_* \downarrow \mathbf{Sign}^{()})$, and it is cocomplete by Lemma 8.4.5. By Lemma 8.4.4, $INTO(\mathbf{Sign}^{R()*})$ is complete. By the dual of Lemma 8.4.5, $(\mathbf{1} \downarrow Proj)$ is complete. □

Theorem 8.4.7 **Parch** is complete.

Proof:
Let *Apply* be the functor indicated by the following diagram:

8.4. Putting parchments together using limits

$$(\mathbf{1} \downarrow Proj) \xrightarrow{Apply} \mathbf{Sign}^{R()*^{op}} \xrightarrow{\mathbf{Mod}^{R()*}} \mathbf{CAT}$$

$$\begin{array}{ccc}
(\Gamma, Lang) & \mapsto & Lang\,\Gamma \\
(\Phi, \alpha) \downarrow & & \uparrow \alpha_\Gamma \\
(\Gamma', Lang') & \mapsto & Lang'\,\Gamma' = Lang'\,\Phi\,\Gamma
\end{array}$$

$\mathbf{Mod}^{R()*} \circ Apply: ((\mathbf{1} \downarrow Proj)^{op})^{op} \longrightarrow \mathbf{CAT}$ is an indexed category with $Flat(\mathbf{Mod}^{R()*} \circ Apply)$ equivalent to $\mathbf{Parch}^{op}$.

Now $(\mathbf{1} \downarrow Proj)^{op}$ is cocomplete by the dual of Lemma 8.4.6. $\mathbf{Mod}^{R()*}\,Apply(\Gamma, Lang) = \mathbf{Mod}^{R()*}\,Lang\,\Gamma$ is cocomplete by Lemma 8.4.2, and $\mathbf{Mod}^{R()*}\,Apply(\Phi, \alpha) = \mathbf{Mod}^{R()*}\,\alpha_\Gamma$ has a left adjoint by Lemma 8.4.3. Thus, by Theorem 8.4.1, $Flat(\mathbf{Mod}^{R()*} \circ Apply) \cong \mathbf{Parch}^{op}$ is cocomplete. □

Since the proof of Theorem 8.4.7 proceeds on a rather abstract level, let us spell out some details of the construction of Theorem 8.4.1 for the case of pullbacks

$$\begin{array}{ccc}
(\mathbf{Sign}', Lang', \Gamma', \mathcal{G}') & \xrightarrow{(\Phi^3, \alpha^3, g^3)} & (\mathbf{Sign}_2, Lang_2, \Gamma_2, \mathcal{G}_2) \\
(\Phi^4, \alpha^4, g^4) \downarrow & & \downarrow (\Phi^2, \alpha^2, g^2) \\
(\mathbf{Sign}_1, Lang_1, \Gamma_1, \mathcal{G}_1) & \xrightarrow{(\Phi^1, \alpha^1, g^1)} & (\mathbf{Sign}, Lang, \Gamma, \mathcal{G})
\end{array}$$

in **Parch**. The first step is to take a pullback in **CAT** ...

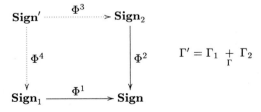

...then take a pushout in $\mathbf{Sign}^{R()*}$ ...

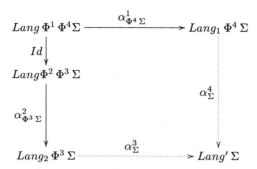

...and finally take a pushout in $\mathbf{Mod}^{R()*} \, Lang' \, \Gamma'$:

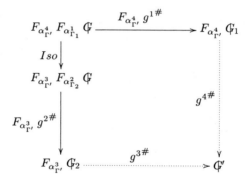

The latter pushout can be constructed as follows: Let

$LangPO = Lang' \, \Gamma'$ then $\alpha^4_{\Gamma'}(Diag(\mathcal{G}_1))$ then $\alpha^3_{\Gamma'}(Diag(\mathcal{G}_2))$ then
$c^{g^1_s(a)} = c^{g^2_s(a)} : \alpha^4_{\Gamma'}(\alpha^1_{\Gamma_1}(s))$ for $a \in \mathcal{G}_s, s \in S(Lang \, \Gamma)$

Then, $\mathcal{G}' = T_{LangPO} \vert_{Lang' \, \Gamma'}$. That is, we combine diagrams of $\mathcal{G}_1$ and $\mathcal{G}_2$, equate the images of $\mathcal{G}$ and interpret the result freely.

With this, we can compute the pullback

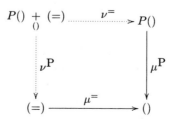

8.4. Putting parchments together using limits

Let $P() \underset{()}{+} (=)$ be $(\mathbf{Sign}, Lang, \Gamma, ⌢\!G)$. Then

- $\mathbf{Sign} = \mathbf{Sign}^{P()}$,
- $\Gamma = \Gamma^{P()}$,
- $Lang\,\Sigma = Lang^{P()}\,\Sigma \cup Lang^{(=)}\,\Sigma$,
- $⌢\!G_{Term(X,s)} = [Env(X) \rightharpoonup s]$,
- $[Env(X) \longrightarrow \mathbf{Bool}\ ] \subset ⌢\!G_{Formula(X)}$ is a proper subset, because $=_{⌢\!G} (f_1, f_2)$ generates new data for partial maps f_1, f_2,
- $\mathbf{Bool} \subset ⌢\!G_*$ is a proper subset, because $bind_{⌢\!G}(=_{⌢\!G} (f_1, f_2))$ generates new data for partial maps f_1, f_2, and
- $D_{⌢\!G} = \{\,true\,\}$.

We are now one step further: there is some interaction of concepts, we have equations over partial operation symbols in $P() \underset{()}{+} (=)$! But their semantical interpretation is a bit funny. The problem is that there is no hint on how to interpret equations over partial operation symbols. Thus the pullback construction generates a "free" interaction of concepts, yielding a "junk" truth value for each possible equation involving partial term evaluation. Though it happens that $I(P() \underset{()}{+} (=))$ is $P(\stackrel{e}{=})$, the presence of so many truth values causes problems when using $P() \underset{()}{+} (=)$ for further combinations. For example, adding negation or implication does not behave in the expected way: a formula containing an undefined term always is non-true (i. e. false, under the functor I), so $t \stackrel{e}{=} t$ and $\neg t \stackrel{e}{=} t$ would both be false in case of undefinedness of t.

The solution to this problem: We have to specify how equality is interpreted on partial term evaluation! This can be done by defining a congruence relation on $⌢\!G$. For example, define a congruence $\stackrel{e}{\equiv}$ on $⌢\!G$ generated by

$$=_{⌢\!G} (f_1, f_2) \stackrel{e}{\equiv} \lambda v.\begin{cases} true, & \text{if } f_1(v), f_2(v) \text{ are defined and equal} \\ false, & \text{otherwise} \end{cases}$$

Then the quotient $⌢\!G \xrightarrow{\_/\stackrel{e}{\equiv}} ⌢\!G/\stackrel{e}{\equiv}$ leads to a parchment morphism

$$P(\stackrel{e}{=}) \xrightarrow{(Id, Id, \_/\stackrel{e}{\equiv})} P() \underset{()}{+} (=)$$

The resulting quotient reduces the junk, and preservation of the truth value space is restored. And $P(\stackrel{e}{=})$ is the parchment of many-sorted partial algebras with existence equations [Bur86]! Note that taking a quotient semantical algebra is a special case of limit of parchments.

Similarly, we can define congruences $\stackrel{w}{=}$ and $\stackrel{s}{=}$ leading to parchments $P(\stackrel{w}{=})$ (weak equations), and $P(\stackrel{s}{=})$ (strong equations, see section 3.2).

We also can get a three-valued logic.

Let $t_1 \stackrel{3}{=} t_2$ be $\begin{cases} true, & \text{if } t_1, t_2 \text{ are defined and equal} \\ false, & \text{if } t_1, t_2 \text{ are defined and not equal} \\ \bot, & \text{otherwise} \end{cases}$

Identify a three-valued map $f: Env(X) \longrightarrow \{true; false; \bot\}$ with $=_{\mathcal{G}} (f_1, f_2)$ for some f_1, f_2 satisfying $f(v) = f_1(v) \stackrel{3}{=} f_2(v)$ and let $\stackrel{3}{=}$ be the congruence relation generated by $=_{\mathcal{G}} (f_1, f_2) \stackrel{3}{=} \lambda v. f_1(v) \stackrel{3}{=} f_2(v)$. Then

$$P(\stackrel{3}{=}) \xrightarrow{(Id, Id, \_/\stackrel{3}{=})} P() \underset{()}{+} (=)$$

yields partial algebras with a three valued logic.

Adding relations can be done using $R(R)$ which is () augmented by relation symbols in the signatures, relations in the models and applications of relation symbols to terms in the sentences.

$$RP(R \stackrel{e}{=}) \xrightarrow{\_/\stackrel{R}{=}} P(\stackrel{e}{=}) \underset{()}{+} R(R) \cdots\cdots\rightarrow P(\stackrel{e}{=})$$
$$\phantom{RP(R \stackrel{e}{=}) \xrightarrow{\_/\stackrel{R}{=}} P(\stackrel{e}{=}) \underset{()}{+}} \Big\downarrow \Big\downarrow$$
$$\phantom{RP(R \stackrel{e}{=}) \xrightarrow{\_/\stackrel{R}{=}} P(\stackrel{e}{=}) \underset{()}{+}} R(R) \longrightarrow ()$$

where $\stackrel{R}{=}$ is the congruence relation generated by

$$R(X)_{\mathcal{G}}(f_1, \ldots, f_n) \stackrel{R}{=} \lambda v. \begin{cases} true, & \text{if } (f_1(v), \ldots, f_n(v)) \text{ def. and } \in R \\ false, & \text{otherwise} \end{cases}$$

Similarly, we get $P(\stackrel{3}{=} R)$.

Adding partiality to first order logic can be done with the parchment $FOL(R =)$ (total first order logic), where universal quantification is modeled by an overloaded operation $\forall(Y): Formula(X \cup Y) \longrightarrow Formula(X)$ (cf. [Ste92]):

8.4. PUTTING PARCHMENTS TOGETHER USING LIMITS

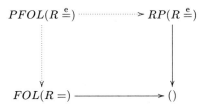

In the three valued case,

$$PFOL(\stackrel{3}{=}R) \xrightarrow{/\stackrel{FOL3}{\equiv}} P(\stackrel{3}{=}R) \underset{()}{+} FOL(R=) \dashrightarrow P(\stackrel{3}{=}R)$$

$$\downarrow \qquad\qquad\qquad\qquad\qquad \downarrow$$

$$FOL(R=) \longrightarrow ()$$

we have to specify a congruence $\stackrel{FOL3}{\equiv}$ defining three-valued logical connectives and quantifiers. For example, we could use Kleene's definitions [Kle52] here, which are also used in VDM [Jon90].

Due to lack of time, we cannot consider parchments for institutions which are built using not only partial term evaluation, but also *partial variable valuations* (see [Sco79, AC96]).

With help of the examples, we are now ready to prove

Proposition 8.4.8 **TruthParch** is *not* complete!

Proof:

Suppose that $(P, \rho^=, \rho^P)$ is a pullback of $(\mu^P, \mu^=)$ in **TruthParch**. By the universal property of $P() \underset{()}{+} (=)$, there exists a unique $\mu: P \longrightarrow P() \underset{()}{+} (=)$ in **Parch** with $\nu^= \circ \mu = \rho^=$ and $\nu^P \circ \mu = \rho^P$. For $x \in \{e, s, w\}$, we have constructed $(Id, Id, g^x): P(\stackrel{x}{=}) \longrightarrow P() \underset{()}{+} (=)$ with $g^x = (\_/\stackrel{x}{\equiv})$ above.

Now $\nu^= \circ (Id, Id, g^x)$ and $\nu^P \circ (Id, Id, g^x)$ both preserve truth. Thus the outer diagram commutes in **TruthParch**. By the universal property of P, there is a $\nu^x: P(\stackrel{x}{=}) \longrightarrow P$. By the universal property of $P() \underset{()}{+} (=)$ it follows that $(Id, Id, g^x) = \mu \circ \nu^x$. Now taking $\mu = (\Phi, \alpha, g)$, $\nu^x = (\Phi^x, \alpha^x, h^x)$, $\mathcal{G}^+ := \mathcal{G}^{P() \underset{()}{+} (=)}$ and $\mathcal{G}^x := \mathcal{G}^{P(\stackrel{x}{=})}$, let $f \in [Env(X) \longrightarrow s]$ be the totally undefined function. Let $a = g_*(bind_{\mathcal{G}^+}(=_{\mathcal{G}^+}(f, f)))$. Then $h^x_*(a) =$

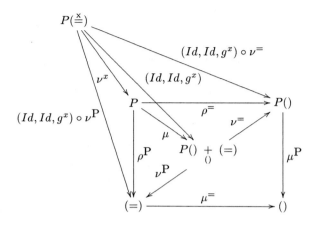

$$= h_*^x(g_*(bind_{\mathcal{G}^+}(=_{\mathcal{G}^+}(f,f))))$$
$$= g^x(bind_{\mathcal{G}^+}(=_{\mathcal{G}^+}(f,f))) = bind_{\mathcal{G}^x}(=_{\mathcal{G}^x}(g^x_{Term(X,s)}(f),g^x_{Term(X,s)}(f)))$$
$$= bind_{\mathcal{G}^x}(=_{\mathcal{G}^x}(f,f)) = \begin{cases} true, & \text{if } x \in \{s,w\} \\ false, & \text{if } x = e \end{cases}$$

Thus, regardless if $a = true$ or $a = false$, h^x cannot preserve truth for all $x \in \{e,s,w\}$, and therefore, $(P, \rho^=, \rho^P)$ is not a pullback in **TruthParch**. □

8.5 Comparison with related work

I have argued that the category of institutions and institution morphisms only allows combinations of institutions *without true interaction*. But Goguen's and Burstall's notion of parchment [GB85] can be equipped with a notion of parchment morphism. The category of parchments turns out to be complete, and parchments combine via limits *with* true interaction.

However, it can happen that this combination introduces new, freely generated truth-values. This indicates the further need for specification of semantical evaluation in the combination, which cannot be automatically inferred from the components in all cases. Instead, the semantical interaction has to be specified explicitly using a congruence relation on the semantical algebra of the parchment, generated by a small number of axioms. These axioms specify how combined things behave semantically.

Thus design decisions are not scattered all over the definitions, but are made at a clearly singled-out point. This can be the basis for a semi-automatic process

8.5. COMPARISON WITH RELATED WORK

of combination of institutions.

These ideas have been illustrated by introducing partiality into different institutions ranging from equational to first-order logic.

Hendrik Hilberdink's work on generating syntax through ends of purely model-theoretic functors [Hil95] has some formal similarity to parchments. In his equation

$$T_\Sigma(X)_s = \int_m Set(Set^{T(S)}(X, \mathcal{E}_{(S,\Sigma)}(m)), \mathcal{E}_{(S,\Sigma)}(m)(s))$$

$Set^{T(S)}(X, \mathcal{E}_{(S,\Sigma)}(m))$ corresponds to our $Env(X)$ and $Set(Set^{T(S)}(X, \mathcal{E}_{(S,\Sigma)}(m)), \mathcal{E}_{(S,\Sigma)}(m)(s))$ corresponds to our $[Env(X) \longrightarrow s]$ (viewed as part of the *Lang m*-reduct of $\mathcal{G}$). But there are two main differences:

1. While the end $\int_m \ldots$ *generates* term syntax and term evaluation from model categories, parchments allow to *specify* term syntax (via the *Lang*-functor) and term evaluation (via the semantical algebra $\mathcal{G}$) in a modular, "global" way leading to the possibility of well-structured combinations. *Generating* term syntax out of model categories does not free us from *specifying* it: In many situations, we cannot cope with the realm of all possible operations consistent with model morphisms, but we have to single out particular ones. It is not clear to me what this difference means for the combination process, since it is even not clear to me how Hilberdink's approach allows combinations and if there is a completeness theorem. But perhaps both approaches are complementary.

2. For institutions of partial algebras, we have to deal with partial variable valuations and term evaluation. In Hilberdink's approach, therefore *Set* has to replaced by a category of partial maps, and it is not clear if the results generalize for this case.

Chapter 9

Conclusion

> A fly and a flea in a flue
> Were imprisoned, so what could they do?
> Said the fly, 'Let us flee',
> 'Let us fly', said the flea.
> So they flew through a flaw in the flue.[1]

To the general theory of institutions and logical systems, this thesis contributes a systematic development of notions of representation that may relax the distinction between signatures and sentences in different degrees:

1. Simple institution representations allow to map signatures to theories, sentences to sentences and models to models. Theorem provers can be re-used along such representations (under some extra technical condition).

2. Conjunctive institution representations are like simple representations, except that a sentence may be translated to a finite set of sentences. Theorem provers can still be re-used.

3. Weak institution representations map theories to theories and models to models. Under some condition, this can be made modular, such that sentences are mapped to theory extensions, and for mapping a theory, we then have to collect all the theory extensions. Semantic properties are kept along such representations, but since sentences are mapped to whole theory extensions, there is only a restricted re-use of theorem provers.

4. The usefulness of institution semi-representations is not so clear, perhaps there is a useful link to institution semi-morphisms

---

[1] From J. Dahl: 99 Limericks. Langewiesche-Brandt 1962.

Each notion can be restricted to the case of embeddings; see Chapter 4.

These notions can be introduced using a basic notion of (plain) representation and different monadic constructions that act on the target institution. As a side-effect, the resulting categories of institutions are linked with adjoint functors, and also the relation of institutions to specification frames (which collapse signatures and sentences into theories) has been clarified.

That these abstract notions are useful is then proved in some quasi "empirical" study. With the help of the first three of the above notions of institution representation, we can draw a detailed picture of the relations between various well-known institutions of total, order-sorted and partial algebras. One main result states that the following institutions are weakly equivalently expressive and that there are plenty of simple and conjunctive embeddings among them (see Chapter 5):

1. Relational Partial Conditional Existence-Equational Logic $RP(R \stackrel{e}{=} \Rightarrow R \stackrel{e}{=})$ [Bur82]

2. Partial Conditional Existence Equational Logic $P(\stackrel{e}{=} \Rightarrow \stackrel{e}{=})$ [Bur82, Bur86]

3. Partial Existentially-Conditioned Existence-Equational Logic $P(D \Rightarrow \stackrel{e}{=})$ [Bur82, Jar88, Jar93]

4. Partial Logic With Strong Equations $P(\stackrel{s}{=})$ [Bur82, Hoe81, Kle52, Slo68]

5. Hierarchical Equationally Partial Theories $HEP(\stackrel{e}{=} \Rightarrow \stackrel{e}{=})$, $HEP1(\stackrel{e}{=} \Rightarrow \stackrel{w}{=})$, $HEP1(\stackrel{w}{=})$ [Fre72, Rei87]

6. Limit Theories $R(R =\Rightarrow \exists! R =)$ [Cos79]

7. Left Exact Sketches $LESKETCH$ [BW85, Gra87]

8. Coherent Order Sorted Algebras With Sort Constraints $COS(=:\Rightarrow=:)$ [MG93]

These form the strongest level of a hierarchy of expressiveness of institutions consisting altogether of five levels. The hierarchy is strict at the semantical level. The levels of the hierarchy correspond to different degrees of conditionality that may be used in the axioms, namely conditional statements about equations, about relations, and about data (Chapter 6).

If the notion of embedding is weakened in such a way that free constructions and (under an extra assumption) proof theory still are preserved (technically, we get categorical retractive representations), then the picture changes. There seem to be only two levels of expressiveness: conditional axioms and unconditional axioms. At the level of conditional axioms, there are special restrictions of well-known institutions that have some intended pragmatic meaning of their theories and models. It turns out that this intended pragmatic meaning can be formalized as a categorical retractive representation, and this representation captures the essential difference of two institutions which are equivalent w. r. t. categorical retractive representations but not w. r. t. embeddings.

These results help to unify the different branches of specification of partial algebras from a semantical point of view. Further, one can combine specifications written in different logical systems by multi-paradigm specification [AC94] using the embeddings. Along the many simple and conjunctive institution representations set up in this work, entailment systems and theorem provers can be re-used [CM93].

The institution graphs set up in chapter 5 should allow to navigate among different logical systems. Perhaps an interactive system might find out the weakest expressive logical system in which the specification under consideration can be expressed, and then point out which edges of the graphs may be used for translation to other logical systems.

Future research should pass over to the examination of institutions and logics with stronger or incomparable expressiveness, which capture other concepts like concurrency or object-orientation. For this, it is essential to be able to talk about and integrate different concepts simultaneously. The notation for institutions used throughout this work suggests that institutions are built in a modular manner. Chapter 8 is a starting point in the examination of the mathematical foundations for such combinations. Combinations of representations, studied in [MTP98], would depart from the *ad-hoc* manner of the construction of representations and make a science of it. But until the representation and combination of logical systems is fully understood, a lot of further work has to be done.

Appendix A

Some category-theoretic preliminaries

> "Perhaps the purpose of categorical algebra is to show that which is trivial is trivially trivial." *Peter Freyd*[1]

Throughout this thesis, familiarity with category-theoretic notions such as category, functor, natural transformation, limit, adjoint is assumed. Readers who are not familiar with this may consult any textbook, for example, [AHR88] or [Bor94]. In this appendix, some category-theoretic notions that may be not so common are recalled.

A.1 Monads and Kleisli categories

A *monad* on a category $\mathbf{X}$ is a triple $\mathbf{T}=(T, \eta, \xi)$ where $T: \mathbf{X} \longrightarrow \mathbf{X}$ is a functor and $\eta: 1_{\mathbf{X}} \longrightarrow T$ and $\xi: T \circ T \longrightarrow T$ are natural transformations satisfying the commutativity conditions

$$\xi \circ (\eta_T) = 1_T = \xi \circ (T\,\eta) \quad \textbf{identity}$$

$$\xi \circ (\xi_T) = \xi \circ (T\,\xi) \quad \textbf{associativity}$$

A monad $\mathbf{T}$ induces the Kleisli category $\mathbf{X_T}$. Our intuition behind it is that its morphisms go into some "enriched" target and thus can be constructed more flexibly than morphisms of $\mathbf{X}$. $\mathbf{X_T}$ is defined by:

---
[1] Proceedings of the Conference on Categorical Algebra. Cited from [HS73].

- the objects of $\mathbf{X_T}$ are those of $\mathbf{X}$
- a morphism $f\colon X \longrightarrow Y$ in $\mathbf{X_T}$ is a morphism $f\colon X \longrightarrow T(Y)$ in $\mathbf{X}$
- the composite of $f\colon X \longrightarrow Y$ and $g\colon Y \longrightarrow Z$ is given in $\mathbf{X}$ by the composite
$$X \xrightarrow{f} T(Y) \xrightarrow{T(g)} T(T(Z)) \xrightarrow{\xi_Z} T(Z)$$
- the identity on an object X of $\mathbf{X_T}$ is just $\eta_X\colon X \longrightarrow T(X)$ in $\mathbf{X}$

Moreover, there is a pair of adjoint functors between $\mathbf{X}$ and $\mathbf{X_T}$. $U_{\mathbf{T}}(X \xrightarrow{F} T(Y))$ is $T(X) \xrightarrow{T(f)} T(T(Y)) \xrightarrow{\xi_Y} T(Y)$, and $F_{\mathbf{T}}(X \xrightarrow{f} Y)$ is $X \xrightarrow{f} Y \xrightarrow{\eta_Y} T(Y)$. The unit of the adjunction is just η.

Given any adjoint situation $(\zeta, \epsilon)\colon F \dashv U\colon \mathbf{A} \longrightarrow \mathbf{X}$, putting $T = U \circ F$, $\xi = U\zeta_F$ yields a monad (T, ζ, ξ) on $\mathbf{X}$. Moreover, there is a full and faithful comparison functor $K_{\mathbf{T}}\colon \mathbf{X_T} \longrightarrow \mathbf{A}$, comparing the Kleisli adjunction of the monad with the given adjunction. It is given by $K_{\mathbf{T}}(X \xrightarrow{f} T(Y)) = F(X) \xrightarrow{F(f)} F(U(F(Y))) \xrightarrow{\epsilon_{F(Y)}} F(Y) = F(X) \xrightarrow{f^\#} F(Y)$.

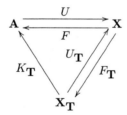

A.2 Multiple pushouts, multiple pullbacks and amalgamation

Let $\mathcal{S} = (A, (A \xrightarrow{g_i} B_i)_{i \in I})$ be a source (that is, a class of arrows with common domain. Note that the domain must be given separately because I may be empty). A colimit of the diagram consisting of all objects and arrows of the source is called a *multiple pushout* of the source. It is denoted by $Colim(\mathcal{S}) = (C, A \xrightarrow{\rho} C, (B_i \xrightarrow{\rho_i} C)_{i \in I})$, where $C = tip(Colim(\mathcal{S}))$ is the tip of the colimiting cocone, while ρ and the ρ_i are the injections. A category with *canonical* multiple pushouts is a category endowed with a distinguished multiple pushout for each source, such that the result of pasting together distinguished pushouts again is a distinguished pushout.

The dual notion is *multiple pullback* (see [AHS90, 11L]). In **CAT** multiple pullbacks are construction via *amalgamation*: Let $((\mathbf{A}_i \xrightarrow{G_i} \mathbf{B})_{i \in I}, \mathbf{B})$ be a sink in **CAT** and $(\mathbf{A}, (\mathbf{A} \xrightarrow{F_i} \mathbf{A}_i)_{i \in I}, \mathbf{A} \xrightarrow{F} \mathbf{B})$ be its multiple pullback. Then the objects of **A** can be written as amalgamated sums $\underset{B}{+} (A_i)_I$, where $A_i \in |\mathbf{A}_i|$ with $G_i(A_i) = B$ for a fixed object $B \in \mathbf{B}$, and $\underset{B}{+} (A_i)_I$ is the unique object with $F_i(\underset{B}{+} (A_i)_I) = A_i$ for all $i \in I$. For the morphisms, a similar relation holds.

Definition A.2.1 A full subcategory **A** of **B** is called

- *colimit-closed*, if each colimit in **B** of a diagram in **A** already belongs to **A** (and, by fullness of the subcategory, is a colimit in **A** of the diagram as well);

- *isomorphism-closed* provided that every **B**-object that is isomorphic to some **A**-object is itself an **A**-object;

- *isomorphism-dense* provided that every **B**-object is isomorphic to some **A**-object. □

Proposition A.2.2 A full coreflective subcategory **A** of **B** is colimit-closed in **B** if and only if **A** is isomorphism-closed in **B**.

Proof:
Take the dual of Proposition 13.27 in [AHS90]. □

Proposition A.2.3 Left adjoint functors preserve colimits.

Proof:
Dualize Proposition 18.9 in [AHS90]. □

A.3 Locally finitely presentable categories

Definition A.3.1 ([AR94]) An object K of a category $\mathcal{K}$ is called *finitely presentable* provided that its hom-functor

$$hom(K, \_) : \mathcal{K} \longrightarrow \mathbf{Set}$$

preserves directed colimits.

For example, a set is finitely presentable in **Set** iff it is finite. A many-sorted algebra is finitely presentable in $\mathbf{Mod}(S, OP)$ iff it can be presented by finitely many generators and finitely many equations in the usual algebraic sense.

Definition A.3.2 ([AR94]) A category $\mathcal{K}$ is called *locally finitely presentable* provided that it is cocomplete and has a set $\mathcal{A}$ of finitely presentable objects such that every object is a directed colimit of objects from $\mathcal{A}$.

Proposition A.3.3 ([AR94]) Each locally finitely presentable category is complete. □

Locally finitely presentable categories are categories that satisfy some completeness properties (completeness and cocompleteness) and some smallness properties (roughly speaking, they are categories of structures with operations of finite arities).

A.4 Effective equivalence relations

Definition A.4.1 (cf. [AR94, 3.4(8)])

A *relation* on an object K is a subobject of $K \times K$ (usually represented by a pair $e_1, e_2 \colon E \longrightarrow K$ of morphisms such that the morphism $\langle e_1, e_2 \rangle \colon E \longrightarrow K \times K$ is a monomorphism). We call $e_1, e_2 \colon E \longrightarrow K$ an *equivalence relation* provided that it is

1. *reflexive*, i. e. , the diagonal of $K \times K$ is contained in the subobject represented by $\langle e_1, e_2 \rangle$

2. *symmetric*, i. e. , the monomorphisms $\langle e_1, e_2 \rangle$ and $\langle e_2, e_1 \rangle$ represent the same subobject

A.5. FOUNDATIONAL ISSUES

3. *transitive*, i. e., when we form the pullback of e_2 and e_1:

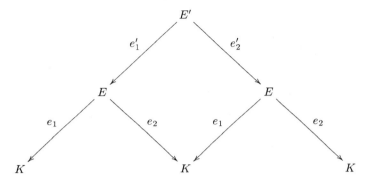

then the subobject represented by $\langle e_1 \circ e'_1, e_2 \circ e'_2 \rangle$ is contained in that represented by $\langle e_1, e_2 \rangle$.

Definition A.4.2 A *kernel pair* of a morphism $f \colon K \longrightarrow K'$ in a category is a pair $e_1, e_2 \colon E \longrightarrow K$ such that

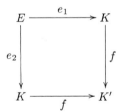

is a pullback.

A category has *effective equivalence relations*, if every equivalence relation is the kernel pair of some morphism.

A.5 Foundational issues

For the theory of logical systems, there is the problem that the standard foundations of mathematics (for example, Zermelo-Fraenkel set theory with axiom of choice ZFC) does not suffice: all standard institutions have collections of models for a given signature that are proper classes (resp. categories that are not small, i.e. have a proper class of objects).

This problem already arises within category theory and is solved as follows. Extend ZFC to $ZFCU$, where $ZFCU$ contains additional axioms stating that there is a set U ("universe"), which is an inner model of ZFC (i.e. closed under all standard set-theoretic operations such as products and powersets). Now call the "sets" in $ZFCU$ conglomerates, the subconglomerates of U classes and the elements of U sets. Then a category has to have classes of objects and morphisms, while a quasicategory can have arbitrary conglomerates of objects and morphisms. Likewise, there are functors and quasifunctors [HS73].

The existence of a universe U is essentially equivalent to the existence of a strongly inaccessible cardinal. It is impossible to prove the relative consistency of this axiom. This is because we can derive

$$ZFCU \vdash Con(ZFC) \qquad \text{(A.1)}$$

because U is an inner model of ZFC, so ZFC can proved to be consistent in $ZFCU$. The situation is totally different from the Axiom of Choice or the Continuum Hypotheses. Gödel proved the relative consistency of both, that is, for the Axiom of Choice,

$$ZF \vdash Con(ZF) \Rightarrow Con(ZFC) \qquad \text{(A.2)}$$

so the Axiom of Choice cannot introduce any inconsistencies which are not already present in ZF. But if we could prove

$$ZFC \vdash Con(ZFC) \Rightarrow Con(ZFCU) \qquad \text{(A.3)}$$

then together with (A.1) we get

$$ZFCU \vdash Con(ZFCU) \qquad \text{(A.4)}$$

from which, by the second Gödel incompleteness theorem, the inconsistency of $ZFCU$ follows.

But in spite of this, generally $ZFCU$ is felt to be free of inconsistencies. Usually only one universe is needed, but there is the possibility of assuming an infinite hierarchy of universes (Grothendieck), with even more possibilities of inconsistencies.

Returning to logical systems, for institutions, the component **Mod** is a quasifunctor from the category of signatures to the quasicategory of all categories. Quasicategories and quasifunctors behave in many respects like categories and functors. But there are differences, because category theory often deals with smallness conditions. For example, categories have usually only limits of set-indexed diagrams, while limits of class-indexed diagrams only exist in thin cat-

A.5. FOUNDATIONAL ISSUES

egories (= pre-orders). So, limits in non-thin quasicategories only can exist if they are indexed by not too large conglomerates.

The essential thing to prove is that **PlainInst**, **Inst**, **Ins** and so on are quasicategories. This is not entirely trivial, because the objects of those categories, institutions, contain quasifunctors as components. But both the domain and the codomain of those quasifunctors are bounded by a fixed conglomerate.

The categorical logics of Meseguer [Mes89] are institutions (or logics) only if models are small categories. But this leads to the funny possibility of embedding many of those categorical logics into $P(\stackrel{e}{=}\Rightarrow\stackrel{e}{=})$ (partial algebras with conditional existence equations) by a jump between meta and object level, see section 5.2. This jump is not possible if models are large categories (i.e. with a class of objects), but then the resulting categorical logic lives in a higher Grothendieck universe.

Another need for higher Grothendieck universes is **Parch**, due to its large objects Γ and $\mathcal{G}$. In [MTP98], **Parch** is replaced by a similar category, which actually is a quasicategory, so no higher Grothendieck universe is needed any more.

Bibliography

[AHR88] J. Adámek, H. Herrlich, J. Rosický. Essentially equational categories. *Cahiers Topologie Géom. Differentielle* **29**, 175–192, 1988.

[AHS90] J. Adámek, H. Herrlich, G. Strecker. *Abstract and Concrete Categories*. Wiley, New York, 1990.

[AR94] J. Adámek, J. Rosický. *Locally Presentable and Accessible Categories*. Cambridge University Press, 1994.

[AD92a] V. Antimirov, A. Degtyarev. Consistency and semantics of equational definitions over predefined algebras. In M. Rusinowitch, J.-L. Rémy, eds., *Conditional Term Rewriting Systems, Third International Workshop, Lecture Notes in Computer Science* **656**, 67–81. Springer Verlag, 1992.

[AD92b] V. Antimirov, A. Degtyarev. Consistency of equational enrichments. In A. Voronkov, ed., *Logic Programming and Automated Reasoning 92, Lecture Notes in Computer Science* **624**, 313–402. Springer Verlag, 1992.

[AF96] M. Arrais, J. L. Fiadeiro. Unifying theories in different institutions. In M. Haveraaen, O. Owe, O.-J. Dahl, eds., *Recent Trends in Data Type Specifications. 11th Workshop on Specification of Abstract Data Types, Lecture Notes in Computer Science* **1130**, 81–101. Springer Verlag, 1996.

[AC92] E. Astesiano, M. Cerioli. Relationships between logical frameworks. In M. Bidoit, C. Choppy, eds., *Proc. 8th ADT workshop, Lecture Notes in Computer Science* **655**, 126–143. Springer Verlag, 1992.

[AC94] E. Astesiano, M. Cerioli. Multiparadigm specification languages: a first attempt at foundations. In C.M.D.J. Andrews, J.F. Groote, eds., *Semantics of Specification Languages (SoSl 93)*, Workshops in Computing, 168–185. Springer Verlag, 1994.

[AC96] E. Astesiano, M. Cerioli. Non-strict don't care algebras and specifications. *Mathematical Structures in Computer Science* **6**, 85–125, 1996.

[Bal88] J. T. Baldwin. *Fundamentals of Stability Theory*. Springer Verlag, 1988.

[Bar89] M. Barr. Models of Horn theories. *Contemporary Mathematics* **92**, 1–7, 1989.

[BW85] M. Barr, C. Wells. *Toposes, Triples and Theories, Grundlehren der mathematischen Wissenschaften* **278**. Springer Verlag, 1985.

[Bau91]   H. Baumeister. Unifying initial and loose semantics of parameterized specifications in an arbitrary institution. In S. Abramsky, T.S.E. Maibaum, eds., *TAPSOFT 91 Vol. 1: CAAP 91, Lecture Notes in Computer Science* **493**, 103–120. Springer Verlag, 1991.

[BT87]   J. A. Bergstra, J. V. Tucker. Algebraic specifications of computable and semicomputable data types. *Theoretical Computer Science* **50**, 137–181, 1987.

[BW85]   S. L. Bloom, E. G. Wagner. *Many-sorted theories and their algebras with some applications to data types*, 133–168. Cambridge University Press, 1985.

[Bor94]   F. Borceux. *Handbook of Categorical Algebra I – III*. Cambridge University Press, 1994.

[Bur82]   P. Burmeister. Partial algebras — survey of a unifying approach towards a two-valued model theory for partial algebras. *Algebra Universalis* **15**, 306–358, 1982.

[Bur86]   P. Burmeister. *A model theoretic approach to partial algebras*. Akademie Verlag, Berlin, 1986.

[BG77]   R. M. Burstall, J. A. Goguen. Putting theories together to make specifications. In *Proceedings of the 5th International Joint Conference on Artificial Intelligence*, 1045–1058. Cambridge, 1977.

[Cer93]   M. Cerioli. *Relationships between Logical Formalisms*. PhD thesis, TD-4/93, Università di Pisa-Genova-Udine, 1993.

[CM93]   M. Cerioli, J. Meseguer. May I borrow your logic? In A.M. Borzyszkowski, S.Sokolowski, eds., *Proc. MFCS'93 (Mathematical Foundations of Computer Science), Lecture Notes in Computer Science* **711**, 342–351. Springer Verlag, Berlin, 1993. Journal version appeared in Theoretical Computer Science **173**, 311–347, 1997.

[CMR99]   M. Cerioli, T. Mossakowski, H. Reichel. From total equational to partial first order logic. In E. Astesiano, H.-J. Kreowski, B. Krieg-Brückner, eds., *Algebraic Foundations of Systems Specifications*, 31–104. Springer Verlag, 1999.

[CGRW95]   I. Claßen, M. Große-Rhode, U. Wolter. Categorical concepts for parameterized partial specification. *Mathematical Structures in Computer Science* **5**, 153–188, 1995.

[CoFI]   CoFI. The Common Framework Initiative for algebraic specification and development, electronic archives. Notes and Documents accessible from http://www.cofi.info.

[CASL]   CoFI Language Design Task Group. CASL – The CoFI Algebraic Specification Language – Summary. Documents/CASL/Summary, in [CoFI], Mar. 2001.

[Cos79]   M. Coste. Localisation, spectra and sheaf representation. In M.P. Fourman, C.J. Mulvey, D.S. Scott, eds., *Application of Sheaves, Lecture Notes in Mathematics* **753**, 212–238. Springer Verlag, 1979.

BIBLIOGRAPHY 191

[Dia01]   R. Diaconescu. Grothendieck institutions. *Applied categorical structures.* to appear.

[EBCO92] H. Ehrig, M. Baldamus, F. Cornelius, F. Orejas. Theory of algebraic module specification including behavioural semantics and constraints. In M. Nivat, C. M. I. Rattray, T. Rus, G. Scollo, eds., *Proc. AMAST 91*, Workshops in Computing, 145–172. Springer Verlag, 1992.

[EM85]   H. Ehrig, B. Mahr. *Fundamentals of Algebraic Specification 1.* Springer Verlag, Heidelberg, 1985.

[EM90]   H. Ehrig, B. Mahr. *Fundamentals of Algebraic Specification 2.* Springer Verlag, Heidelberg, 1990.

[EPO89]   H. Ehrig, P. Pepper, F. Orejas. On recent trends in algebraic specification. In *Proc. ICALP'89, Lecture Notes in Computer Science* **372**, 263–288. Springer Verlag, 1989.

[Far91]   W. A. Farmer. A partial functions version of Church's simple type theory. *Journal of Symbolic Logic* **55**, 1269–1291, 1991.

[Fef92]   S. Feferman. A new approach to abstract data types, I informal development. *Mathematical Structures in Computer Science* **2**, 193–229, 1992.

[FS88]   J. Fiadeiro, A. Sernadas. Structuring theories on consequence. In D. Sannella, A. Tarlecki, eds., *Recent Trends in Data Type Specification, 5th Workshop on Specification of Abstract Data Types, Lecture Notes in Computer Science* **332**, 44–72. Springer Verlag, 1988.

[Fre72]   P. Freyd. Aspects of topoi. *Bull. Austral. Math. Soc.* **7**, 1–76, 1972.

[GU71]   P. Gabriel, F. Ulmer. *Lokal präsentierbare Kategorien, Lecture Notes in Mathematics* **221**. Springer Verlag, Heidelberg, 1971.

[GJM85]   J. Goguen, J.-P. Jouannaud, José Meseguer. Operational semantics of order-sorted algebra. In W. Brauer, ed., *Proceedings, 1985 International Conference on Automata, Languages and Programming, Lecture Notes in Computer Science* **194**, 221–231. Springer, 1985.

[Gog91]   J. A. Goguen. A categorical manifesto. *Mathematical Structures in Computer Science* **1**, 49–67, 1991.

[GB85]   J. A. Goguen, R. M. Burstall. A study in the foundations of programming methodology: Specifications, institutions, charters and parchments. In D. Pitt et al., ed., *Category Theory and Computer Programming, Lecture Notes in Computer Science* **240**, 313–333. Springer Verlag, 1985.

[GB92]   J. A. Goguen, R. M. Burstall. Institutions: Abstract model theory for specification and programming. *Journal of the Association for Computing Machinery* **39**, 95–146, 1992. Predecessor in: LNCS 164, 221–256, 1984.

[GM86]   J. A. Goguen, J. Meseguer. Eqlog: Equality, types, and generic modules for logic programming. In D. DeGroot, G. Lindstrom, eds., *Logic Programming. Functions, Relations and Equations*, 295–363. Prentice-Hall, Englewood Cliffs, New Jersey, 1986.

[GM92] J. A. Goguen, J. Meseguer. Order-sorted algebra I: equational deduction for multiple inheritance, overloading, exceptions and partial operations. *Theoretical Computer Science* **105**, 217–273, 1992.

[GTW78] J. A. Goguen, J. W. Thatcher, E. G. Wagner. An initial algebra approach to the specification, correctness and implementation of abstract data types. In R. Yeh, ed., *Current Trends in Programming Methodology*, 4, 80–144. Prentice Hall, 1978.

[GW88] J. A. Goguen, T. Winkler. Introducing OBJ3. Research report SRI-CSL-88-9, SRI International, 1988.

[Gra87] J. W. Gray. Categorical aspects of data type constructors. *Theoretical Computer Science* **50**, 103–135, 1987.

[GH86] J. V. Guttag, J. J. Horning. Report on the Larch shared language. *Science of Computer Programming* **6**(2), 103–134, 1986.

[HHP93] R. Harper, F. Honsell, G. D. Plotkin. A framework for defining logics. *Journal of the Association for Computing Machinery* **40**, 143–184, 1993.

[HN96] A.E. Haxthausen, F. Nickl. Pushouts of order-sorted algebraic specifications. In *Proceedings of AMAST'96, Lecture Notes in Computer Science* **1101**, 132–147. Springer-Verlag, 1996.

[HS73] H. Herrlich, G. Strecker. *Category Theory*. Allyn and Bacon, Boston, 1973.

[Hil95] H. Hilberdink. The end of syntax. The end of algebra. Manuscripts, St. Peter's College, Oxford, 1995.

[Hoe81] H.-J. Hoehnke. On partial algebras. In *Universal Algebra (Proc. Coll. Esztergom 1977), Colloq. Math. Soc. J. Bolyai* **29**, 373–412. North Holland, Amsterdam, 1981.

[Jar88] G. Jarzembski. Weak varieties of partial algebras. *Algebra Universalis* **25**, 247–262, 1988.

[Jar93] G. Jarzembski. Programs in partial algebras. *Theoretical Computer Science* **115**, 131–149, 1993.

[Jon90] C. B. Jones. *Systematic Software Development Using VDM*. Prentice Hall, 1990.

[Kle52] S.C. Kleene. *Introduction to Metamathematics*. North Holland, 1952.

[Kre87] H.-J. Kreowski. Partial algebras flow from algebraic specifications. In T. Ottmann, ed., *Proc. ICALP 87, Lecture Notes in Computer Science* **267**, 521–530. Springer Verlag, 1987.

[KM95] H.-J. Kreowski, T. Mossakowski. Equivalence and difference of institutions: Simulating horn clause logic with based algebras. *Mathematical Structures in Computer Science* **5**, 189–215, 1995.

[LS86] J. Lambek, P. J. Scott. *Introduction to Higher-Order Categorical Logic*. Cambridge Studies in Advanced Math. Cambridge University Press, 1986.

[Lan72] S. Mac Lane. *Categories for the working mathematician*. Springer, 1972.

BIBLIOGRAPHY 193

[Lei87] D. Leivant. Characterization of complexity classes in higher-order logic. In *Proc. 2nd Annual Conference Structure in Complexity Theory*, 203–217. IEEE, 1987.

[Llo87] J.W. Lloyd. *Foundations of Logic Programming*. Springer Verlag, 1987.

[MSS90] V. Manca, A. Salibra, G. Scollo. Equational type logic. *Theoretical Computer Science* **77**, 131–159, 1990.

[MSS91] V. Manca, A. Salibra, G. Scollo. On the expressiveness of equational type logic. In C. Rattray, R. Clark, eds., *The Unified Computation Laboratory*, 85–100. Oxford University Press, 1991.

[MOM95] N. Martí-Oliet, J. Meseguer. From abstract data types to logical frameworks. In E. Astesiano, G. Reggio, A. Tarlecki, eds., *Recent Trends in Data Type Specification. Proceedings, Lecture Notes in Computer Science* **906**, 48–80. Springer Verlag, London, 1995.

[May85] B. Mayoh. Galleries and institutions. Report DAIMI PB-191, Aarhus University, 1985.

[Mes89] J. Meseguer. General logics. In *Logic Colloquium 87*, 275–329. North Holland, 1989.

[Mes92] J. Meseguer. Conditional rewriting as a unified model of concurrency. *Theoretical Computer Science* **96**(1), 73–156, 1992.

[MG93] J. Meseguer, J. Goguen. Order-sorted algebra solves the constructor, selector, multiple representation and coercion problems. *Information and Computation* **103**(1), 114–158, March 1993.

[Mosa] T. Mossakowski. Parameterized recursion theory – a tool for the systematic classification of specification methods. *Theoretical Computer Science*. To appear.

[Mosb] T. Mossakowski. Relating CASL with other specification languages: the institution level. *Theoretical Computer Science*. July 2003, to appear.

[Mos93] T. Mossakowski. Simulations between various institutions of partial and total algebras. Talk at the Workshop of the ESPRIT Basic Research Working Group COMPASS, Dresden, September 1993.

[Mos95] T. Mossakowski. A hierarchy of institutions separated by properties of parameterized abstract data types. In E. Astesiano, G. Reggio, A. Tarlecki, eds., *Recent Trends in Data Type Specification. Proceedings, Lecture Notes in Computer Science* **906**, 389–405. Springer Verlag, London, 1995.

[Mos96a] T. Mossakowski. Different types of arrow between logical frameworks. In F. Meyer auf der Heide, B. Monien, eds., *Proc. ICALP 96, Lecture Notes in Computer Science* **1099**, 158–169. Springer Verlag, 1996.

[Mos96b] T. Mossakowski. Equivalences among various logical frameworks of partial algebras. In H. Kleine Büning, ed., *Computer Science Logic. 9th Workshop, CSL'95. Paderborn, Germany, September 1995, Selected Papers, Lecture Notes in Computer Science* **1092**, 403–433. Springer Verlag, 1996.

[Mos96c]   T. Mossakowski. Using limits of parchments to systematically construct institutions of partial algebras. In M. Haveraaen, O. Owe, O.-J. Dahl, eds., *Recent Trends in Data Type Specifications. 11th Workshop on Specification of Abstract Data Types, Lecture Notes in Computer Science* **1130**, 379–393. Springer Verlag, 1996.

[Mos98]   T. Mossakowski. Colimits of order-sorted specifications. In F. Parisi Presicce, ed., *Recent trends in algebraic development techniques. Proc. 12th International Workshop, Lecture Notes in Computer Science* **1376**, 316–332. Springer, 1998.

[MTP98]   T. Mossakowski, A. Tarlecki, W. Pawłowski. Combining and representing logical systems using model-theoretic parchments. In F. Parisi Presicce, ed., *Recent trends in algebraic development techniques. Proc. 12th International Workshop, Lecture Notes in Computer Science* **1376**, 349–364. Springer, 1998.

[Mos89]   P. D. Mosses. Unified algebras and institutions. Proceedings of the 4th Annual IEEE Symposium on Logic in Computer Science, 304–312. 1989.

[Mos97]   P. D. Mosses. CoFI: The Common Framework Initiative for Algebraic Specification and Development. In *TAPSOFT '97, Proc. Intl. Symp. on Theory and Practice of Software Development*, volume 1214 of *LNCS*, pages 115–137. Springer-Verlag, 1997.

[Pad88]   P. Padawitz. *Computing in Horn Clause Theories*. Springer Verlag, Heidelberg, 1988.

[Paw96]   W. Pawłowski. Context institutions. In M. Haveraaen, O. Owe, O.-J. Dahl, eds., *Recent Trends in Data Type Specifications. 11th Workshop on Specification of Abstract Data Types, Lecture Notes in Computer Science* **1130**, 436–457. Springer Verlag, 1996.

[Poi86]   A. Poigné. Algebra categorically. In D. Pitt et al., ed., *Category Theory and Computer Programming, Lecture Notes in Computer Science* **240**, 76–102. Springer Verlag, 1986.

[Poi89]   A. Poigné. Foundations are rich institutions, but institutions are poor foundations. In H. Ehrig et al., ed., *Categorical Methods in Computer Science, With Aspects from Topology, Lecture Notes in Computer Science* **393**, 82–101. Springer Verlag, 1989.

[Pri79]   G. Priest. The logic of paradox. *Journal of philosophical logic* **8**, 219–241, 1979.

[Rei87]   H. Reichel. *Initial Computability, Algebraic Specifications and Partial Algebras*. Oxford Science Publications, 1987.

[SS92]   A. Salibra, G. Scollo. A soft stairway to institutions. In M. Bidoit, C. Choppy, eds., *Proc. 8th ADT workshop, Lecture Notes in Computer Science* **655**, 310–329. Springer Verlag, 1992.

[SST92]   D. Sannella, S. Sokolowski, A. Tarlecki. Toward formal development of programs from algebraic specifications: Parameterisation revisited. *Acta Informatica* **29**, 689–736, 1992.

[ST] D. Sannella, A. Tarlecki. *Working with multiple logical systems, In: Foundations of Algebraic Specifications and Formal Program Development*, chapter 10. Cambridge University Press, to appear. See http://zls.mimuw.edu.pl/~tarlecki/book/index.html.

[ST88a] D. Sannella, A. Tarlecki. Specifications in an arbitrary institution. *Information and Computation* **76**, 165–210, 1988.

[ST88b] D. Sannella, A. Tarlecki. Toward formal development of programs from algebraic specifications: implementations revisited. *Acta Informatica* **25**, 233–281, 1988.

[ST90] D. Sannella, A. Tarlecki. Extended ML: Past, present and future. In H. Ehrig, K. P. Jantke, F. Orejas, H. Reichel, eds., *Proceedings of Recent Trends in Data Type Specification, Lecture Notes in Computer Science* **534**, 297–322. Springer, 1990.

[ST93] D. Sannella, A. Tarlecki. Algebraic Specification and Formal Methods for Program Development: What are the Real Problems? In G. Rozenberg, A. Salomaa, eds., *Current Trends in Theoretical Computer Science. Essays and Tutorials*, 115–120. World Scientific Series in Computer Science — Vol. 40, 1993.

[ST86] D. T. Sannella, A. Tarlecki. Extended ML: an institution-independent framework for formal program development. In *Proc. Workshop on Category Theory and Computer Programming, Lecture Notes in Computer Science* **240**, 364–389. Springer, 1986.

[Sco79] D. S. Scott. Identity and existence in intuitionistic logic. In M.P. Fourman, C.J. Mulvey, D.S. Scott, eds., *Application of Sheaves, Lecture Notes in Mathematics* **753**, 660–696. Springer Verlag, 1979.

[SS95] A. Sernadas, C. Sernadas. Theory spaces. Research report, DMIST, 1096 Lisboa, Portugal, 1995. Presented at ISCORE'95 and ADT/COMPASS'95.

[She78] S. Shelah. *Classification Theory and the Number of Non-Isomorphic Models, Studies in Logic and the Foundations of Mathematics* **92**. North Holland, 1978.

[Sho67] J.R. Shoenfield. *Mathematical Logic*. Addison-Wesley, Reading, Massachusetts, 1967.

[Slo68] J. Slominski. *Peano-algebras and quasi-algebras, Dissertationes Mathematicae (Rozprawy Mat.)* **62**. 1968.

[Ste92] P. Stefaneas. The first order parchment. Report PRG-TR-16-92, Oxford University Computing Laboratory, 1992.

[Tar85] A. Tarlecki. On the existence of free models in abstract algebraic institutions. *Theoretical Computer Science* **37**, 269–304, 1985.

[Tar86a] A. Tarlecki. Bits and pieces of the theory of institutions. In D. Pitt, S. Abramsky, A. Poigné, D. Rydeheard, eds., *Proc. Intl. Workshop on Category Theory and Computer Programming, Guildford 1985, Lecture Notes in Computer Science* **240**, 334–363. Springer-Verlag, 1986.

[Tar86b]   A. Tarlecki. Quasi-varieties in abstract algebraic institutions. *Journal of Computer and System Sciences* **33**, 333–360, 1986.

[Tar96a]   A. Tarlecki. Moving between logical systems. In M. Haveraaen, O. Owe, O.-J. Dahl, eds., *Recent Trends in Data Type Specifications. 11th Workshop on Specification of Abstract Data Types, Lecture Notes in Computer Science* **1130**, 478–502. Springer Verlag, 1996.

[Tar96b]   A. Tarlecki. Structural properties of some categories of institutions. Technical report, Warsaw University, 1996.

[Tar99]   A. Tarlecki. Towards heterogeneous specifications. In D. Gabbay and M. d. Rijke, editors, *Frontiers of Combining Systems 2, 1998*, Studies in Logic and Computation, pages 337–360. Research Studies Press, 2000.

[TBG91]   A. Tarlecki, R. M. Burstall, J. A. Goguen. Some fundamentals algebraic tools for the semantics of computation. Part 3: Indexed categories. *Theoretical Computer Science* **91**, 239–264, 1991.

[TWW81]   J. W. Thatcher, E. G. Wagner, J. B. Wright. Specification of abstract data types using conditional axioms. Technical Report RC 6214, IBM Yorktown Heights, 1981.

[Wir86]   M. Wirsing. Structured algebraic specifications: A kernel language. *Theoretical Computer Science* **42**, 123–249, 1986.

[Wol95]   U. Wolter. Institutional frames. In *Recent Trends in Data Type Specification. Proceedings, Lecture Notes in Computer Science* **906**, 469–482. Springer Verlag, London, 1995.

[Yan93]   H. Yan. *Theory and Implementation of Sort Constraints for Order Sorted Algebra*. PhD thesis, Oxford University, 1993.

Index

(), 161, 172
(=), 34, 115, 128, 146, 161
(=⇒=), 34, 98–100, 111, 113, 126, 134, 145
(_)$^+$, 24
(_)$^\heartsuit$, 62
$Apply$, 168
B(=⇒=), 134, 150
COS(=), 43, 103, 104
COS(=:), 43, 104
COS(=:⇒=:), 6, 40, 83, 178
COS(=⇒=), 43, 98
$\mathcal{DNS}$, 143
$\mathcal{DTL}$, 143
ETL, 133, 144
$\mathcal{ETL}$, 35
$F_{\mathbf{Ext}}$, 59
$HEP(\stackrel{e}{=}\Rightarrow\stackrel{e}{=})$, 32, 79, 84, 178
$HEP1(=\Rightarrow\stackrel{w}{=})$, 33, 78, 87
$HEP1(\stackrel{e}{=}\Rightarrow\stackrel{e}{=})$, 78
$HEP1(\stackrel{e}{=}\Rightarrow\stackrel{w}{=})$, 33, 78, 178
$HEP1(\stackrel{w}{=})$, 33, 78, 178
I, 166
$INTO(\mathbf{T})$, 168
$K_{\mathbf{Mod}}$, 63
$LESKETCH$, 6, 40, 84, 86, 93, 178
$NCOS$(=:⇒=:), 43, 82, 93
$NMTP(\stackrel{f}{=})$, 146
$NMTRP(R \stackrel{f}{=})$, 146
$P()$, 161
P(=⇒=)($\stackrel{f}{=}$), 100
$P(D \Rightarrow \stackrel{e}{=})$, 32, 80, 178
$P(\stackrel{e}{=})$, 34, 104, 113, 161, 162
$P(\stackrel{e}{=}\Rightarrow\stackrel{e}{=})$, 32, 34, 79, 80, 101, 124, 144, 178
$P(\stackrel{e}{=}\Rightarrow\stackrel{w}{=})(\stackrel{e}{=})$, 100, 101
$P(\stackrel{e}{=}\stackrel{w}{=})$, 34, 99
$P(\stackrel{f}{=})$, 34, 103, 104, 146
$P(\stackrel{f}{=}\Rightarrow\stackrel{f}{=})$, 34, 69, 136, 137, 141, 150
$P(\stackrel{3}{=} R)$, 172
$P(\stackrel{s}{=})$, 32, 80, 91, 178
$PHOL$, 53
$R(R)$, 100, 172
$R(R \Rightarrow \exists!R =)$, 86
$R(R =)$, 34, 162
$R(R \Rightarrow R)$, 34
$R(R \Rightarrow R =)$, 34, 69, 109, 111, 119, 125, 134, 136, 143–145
$R(R \Rightarrow =)(\exists!R =)$, 36, 100
$R(R \Rightarrow \exists!R =)$, 5, 36, 80, 83, 87, 91, 178
RP(=⇒=)($R \stackrel{f}{=}$), 34, 120, 125, 130, 131
$RP(R \stackrel{e}{=})$, 34
$RP(R \stackrel{e}{=}\Rightarrow R \stackrel{e}{=})$, 5, 27, 79, 82, 109, 178
$RP(R \stackrel{f}{=})$, 34, 121, 126, 128, 145, 146
$RP(R \stackrel{f}{=}\Rightarrow R \stackrel{f}{=})$, 34, 119, 124, 130
$\mathcal{UNIFALG}$, 35
$UR(R \Rightarrow R =)$, 143
ZFC, 185
$ZFCU$, 185
Φ^{embed}, 135
α-extension, 17

$\mu^=$, 166
$\mu^{flatten}$, 150
μ^P, 166
$\mu^=$, 161
μ^D, 143
μ^M, 143
μ^P, 144
$\mu^{P=}$, 146
μ^{PART}, 136, 145, 150
$\mu^{chardom}$, 69
μ^{graph}, 70
μ^P, 162
$\mu^{chardom}$, 79, 145
μ^{char}, 145
μ^{ch}, 83, 98
μ^{cr}, 82
$\mu^{flatten}$, 135, 145
μ^{graph}, 119
μ^{hep}, 79
μ^{hs}, 84
μ^{limgra}, 80
μ^{ls}, 91
μ^{pp}, 80
μ^{rs}, 87
μ^{sc}, 86
μ^{se}, 80
μ^{sn}, 93
$\preceq$, 33
ent, 24
$frame$, 16, 17
log, 24
$sign$, 16
Ext − rps, 56
Ext, 58
Instrps, 52
Inst^{amal}, 51
Parch, 166, 187
PlainEnt, 23
PlainInstrps, 16
PlainInstrps^{amal}, 18
PlainInst, 16
PlainInst^{amal}, 18
PlainInst$_{Mod}$, 63
PlainLog, 24
Pres, 17
SemiInst, 63
SpecFram, 14
SpecFram^{amal}, 18
Th$_0$, 16, 17, 19, 22
TruthParch, 167
WeakInstrps, 53
WeakInst, 53
derive!, 57
derive, 54
$\mathcal{C}^{comp}$, 17
NCOSASIG, 45
OSASIG, 44
REGOSASIG, 45
SIG, 44
BINREL, 111
BIN_REL, 125, 128
BOUNDED_ELEM, 129
BOUNDED_STACK_BODY1, 129
BOUNDED_STACK_BODY2, 130
BOUNDED_STACK, 129
BSTACK, 129
FACTOR, 126
FUNC, 126
GRAPH1, 90
GRAPH2, 90
GRAPH, 124
PATHS, 124
REFLEXIVE, 128
TRANSITIVE1, 90
TRANSITIVE2, 90
TRANSITIVE3, 90
TRANSITIVE, 111
TRANS_REL, 125

abstract syntax, 162
ACT ONE, 26
algebra, 28
algebraic specification, 2
amalgamated objects, 161
amalgamation, 18, 46, 57, 75, 183

INDEX

preservation of, 18

based algebra, 134
borrowing, 10, 25, 51, 52, 60, 62, 64, 67, 77, 139, 143, 178
bounded stacks, 129

carrier set, 28
categorical intersection, 39
categorical retractive representation, 138, 140
categorical retractive simulation, 11
category-theoretic classification, 11
classification theory, 11
closed homomorphism, 38
cocomplete, 167
coequalizer, 38, 112
coherent, 41
coherent signature, 41
colimit, 17, 51, 161
colimit-closed, 183
combination
  free, 170
  of institutions, 160, 175
  of logical systems, 9, 160
comma category, 168
composable signatures, 17, 43, 51
  preservation of, 75
Conditional Equational Logic, 34, 111, 112, 134
conditional formula, 30
conditional generation of various things, 131
conditional term rewriting, 26
congruence, 37, 115, 160, 172
conjunctive embedding of institutions, 67
conjunctive institution representation, 52
conjunctive monad, 52
conjunctively equivalently expressive, 67
connected component, 41
consistency, 186

constraint, 26
context institution, 8
coreflection, 46

definite description operator, 53
designated truth values, 163
diagram completion lemma, 39
difference in expressiveness, 148
difference of institutions, 138

effective equivalence relations, 115, 185
embedding, 140, 178
embedding of institutions, 66
  conjunctive, 67
  simple, 67
  weak, 67
embedding of specification frames, 67
end, 175
entailment system, 8, 23
  re-use of, see borrowing
  representation, 23
eps, 15
Eqlog, 34
equation
  existence, 30
  strong, 32
  weak, 34
Equational Logic, 34
Equational Type Logic, 35
equivalence relation, 184
equivalent in expressiveness, 66, 134, 138, 156
equivalently expressive, 66, 178
  conjunctively, 67
  simply, 67
  weakly, 67
essentially algebraic theories, 33
existence equation, 30
expressiveness, 7, 138, 148, 178
extension lemma, 148

factorization of a function over the image, 126

(regular epi, mono)-factorizations, 109
finite model theory, 63
finitely presentable, 183
flat partiality, 34
foundational problems, 163
FP-sketches, 61
free combination, 170
free completion, 137
free construction, 26, 140, 147, 178
free generating constraint, 4
free interaction of concepts, 171
free model, 10
free PADT, 47
full homomorphism, 37
full surjectivity
    preservation of, 120

generated congruence, 37
Grothendieck universes, 187

HEP-theories, 32, 74
hierarchy theorem, 109, 119, 157, 178
homomorphic extension, 30
homomorphism, 29
    closed, 38
    full, 37
Horn Clause Logic, 34, 109, 111, 143
Horn sentence, 27

implementation, 3, 9
indexed category, 167
initial algebra, 162
initial homomorphism, 163
initial model, 10, 139
initial semantics, 117
initiality constraint, 4
institution, 5, 14
    combination of, 160
institution embedding, 9
institution encoding, 7
institution graph, 7, 179
institution morphism, 9, 161

institution representation, 7, 9, 15, *see* embedding of institution
    conjunctive, 52
    simple, 50
    weak, 59
institution semi-morphisms, 9
institution semi-representation, 63
institution-independence, 5
interaction of concepts, 9, 162, 171, 174
    free, 171
interpretable term, 30
Intuitionistic type theory, 95
invertible pre-institution transformation, 66
isomorphism-closed, 183
isomorphism-dense, 183

junk, 45, 46, 136, 137, 160, 171

kernel, 37
kernel pair, 185
kernels
    preservation of, 121
Kleisli category, 50, 181

least sort, 42
least sort parse, 42
left exact sketch, 39, 74
liberal, 26, 47, 77, 106
liberal institution, 6
liberality
    preservation of, 68
limit, 162
limit theories, 36
locally finitely presentable, 46, 74, 77, 184
logic, 9, 24
    multi-valued, 163
    paraconsistent, 163
    rewriting, 105
    three-valued, 163, 172
    two-valued, 28

logic programming, 34
logic representation, 24
logical consequence, 3
logical structure
  re-use of, *see* borrowing
logical system, 9
  combination of, 160
loose semantics, 147, 156
loose specification, 3

map of institutions, *see* institution representation
meta theory, 7
model, 28
model class monad, 63
model expansion, 57
model morphism, 10
model theory, 9
modularity, 3
monad, 9, 181
  conjunctive, 52
  model class, 63
  theory, 50
  theory extension, 56, 58
monotonicity condition, 40
multi-paradigm specification language, 7
multi-valued logic, 163
multiple pullback, 18, 183
  preservation of, 68
multiple pushout, 17, 18, 56, 59, 182
  preservation of, 22, 70

nested partiality, 34
Noetherian ordering, 43
notation for institutions, 33, 159

operation, 28
order-sorted algebra, 41
order-sorted signature, 40
order-sorted term algebra, 41
overloading, 30, 41, 164

P=NP, 63

PADT, 26, 118
paraconsistent logic, 163
Parameterized abstract data type, 118
parameterized abstract data type, 26
parameterized specification, 3
parameterized theory, 26
paramodulation, 26
parchment, 8, 31, 160, 163
parchment morphism, 164
PART construction, 134–157
partial algebra, 28
Partial Conditional Existence Equational Logic, 32
Partial Conditional Logic, 109
Partial Equational Logic, 112
Partial Existentially-Conditioned Existence-Equational Logic, 32
Partial First-Order Logic, 51
Partial Function Higher-Order Logic, 53
partial operation, 32
partial term evaluation, 171
partial variable valuations, 173
paths over a graph, 124
persistent functor, 129
persistently specifiable specification frame representation, 148
plain map of entailment systems, 23
plain map of institutions, 15
pre-institution, 15
pre-institution representation
  weak rps, 53
presentation, 17
preservation
  of amalgamation, 18
  of composable signatures, 51, 75
  of full surjectivity, 120
  of kernels, 121
  of liberality, 68
  of multiple pullbacks, 68

of multiple pushouts, 22, 70
of regular epis, 111
of regularity, 41
of surjectivity, 119
of truth, 167, 172
programming language, 9
Prolog, 34
proof calculus, 9
proof theory, 9
pullback, 24, 25, 39, 64, 167, 169
pushout, 55, 169

quasicategory, 186
quotient, 38, 112
quotient homomorphism, 111

rank, 40
re-use of theorem provers, *see* borrowing
reduct, 4, 29
reflexive, 184
    Making a relation reflexive, 128
regular epis
    preservation of, 111
regular signature, 40
regularity
    preservation of, 41
relation, 28, 184
Relational Partial Conditional Existence-Equational Logic, 5, 27
relative subalgebra, 37
representation, *see* entailment system representation, *see* institution representation, *see* specification frame representation
    categorical retractive, 138, 140
representation condition, 15
retractive specification frame representation, 139
rewriting logic, 105
rps, 15

safe variable, 137

satisfaction, 31
satisfaction condition, 4, 15, 31
semantical evaluation, 162
semantical universe, 162
semi-computable ADT, 117
sentence, 30
sentence translation, 30
separating properties, 132
signature, 2, 28
signature morphism, 4, 28
    preserves regularity, 41
simple embedding of institutions, 67
simple institution representation, 50
simply equivalently expressive, 67
sort, 28
sort constraint, 42
space of truth values, 160, 163
specifiable ADT, 117
specifiable retractive representation, 147
specifiable retractive specification frame representation, 147
specifiable specification frame representation, 147
specification frame, 8, 13
specification frame representation, 14
    persistently specifiable, 148
    retractive, 139
specification language, 2
specification language construct, 67
specification-building operators, 5
stability theory, 11
strong equation, 32
strongly inaccessible cardinal, 186
subinstitution, 66
subobjects commute with coequalizers, 113
surjectivity
    preservation of, 119
symmetric, 184

term, 29, 41

term algebra, 30
theorem of homomorphisms, 109
theorem provers
    re-use of, *see* borrowing
theory, 16
theory extension monad, 56, 58
theory monad, 50
theory morphism, 14, 16
three-valued logic, 163, 172
topos, 95
total operation, 32
transitive, 185
transitive closure, 125
transporting logical structure, 10
truth
    preservation of, 167, 172
truth values
    designated, 163
    space of, 160, 163
two-valued logic, 28

unified algebras, 35, 133
universal institution, 6

valuation, 30
variable system, 29
VDM, 173

weak embedding of institutions, 67
weak equation, 34
weak institution representation, 59
weak rps pre-institution representation, 53
weakly equivalently expressive, 67

Monographs of the Bremen Institute of Safe Systems

ISSN 1435-8611

1. Buth, Bettina / Berghammer, Rudolf / Peleska, Jan (eds.): Tools for System Development and Verification. Workshop, Proceedings, Bremen, Germany, July 1996. ISBN 3-8265-3806-4. Shaker, 1998.
2. Mossakowski, Till: Representations, Hierarchies and Graphs of Institutions. Dissertation, 1996. Revised version: ISBN 3-89722-831-9, Logos, 2001.
3. Cerioli, Maura / Gogolla, Martin / Kirchner, Héléne / Krieg-Brückner, Bernd / Qian, Zhenyu / Wolf, Markus (eds.): Algebraic System Specification and Development: Survey and Annotated Bibliography. 2nd edition, 1997. ISBN 3-8265-4067-0. Shaker, 1998.
4. Wolff, Burkhart: A Generic Calculus of Transformation. Dissertation. 1997. Revised version: ISBN 3-8265-3654-1. Shaker, 1999.
5. Kolyang: HOL-Z, An Integrated Formal Support Environment for Z in Isabelle/HOL. Dissertation, 1997. ISBN 3-8265-4068-9. Shaker, 1998.
6. Fröhlich, Michael: Inkrementelles Graphlayout im Visualisierungssystem daVinci. Dissertation, 1997. ISBN 3-8265-4069-7. Shaker, 1998.
7. Röfer, Thomas: Panoramic Image Processing and Route Navigation. Dissertation, 1998. ISBN 3-8265-4070-0. Shaker, 1998.
8. Schrönen, Michael: Methodology for the Development of Microprocessor-Based Safety-Critical Systems. Dissertation, 1998. ISBN 3-8265-3655-X. Shaker, 1998.
9. Krieg-Brückner, Bernd / Peleska, Jan / Olderog, Ernst-Rüdiger / Balzer, Dietrich / Baer, Alexander: UniForM Workbench, Universelle Entwicklungsumgebung für Formale Methoden; Schlußbericht. ISBN 3-8265-3656-8. Shaker, 1999.
10. Gärtner, Heino: Schematransformationen in objektorientierten Informationssystemen. Dissertation, 1999. ISBN 3-8265-6542-8. Shaker, 1999.
11. Huge, Anne-Kathrin: Ein Ansatz zur Formalisierung objektorientierter Datenbanken auf der Grundlage von ODMG. Dissertation, 1999. ISBN 3-8265-6543-6. Shaker, 2000.
12. Karlsen, Einar: Tool Integration in a Functional Programming Language. Dissertation. 1998. Revised version. Universität Bremen, 1999.
13. Amthor, Peter: Structural Decomposition of Hybrid Systems. Dissertation. Universität Bremen, 2000.
14. Richters, Mark: A Precise Approach to Validating UML Models and OCL Constraints. Dissertation, 2001. ISBN 3-89722-842-4. Logos, 2001.